Selected Problems in Physical Chemistry

Predrag-Peter Ilich

Selected Problems
in Physical Chemistry

Strategies and Interpretations

 Springer

Dr. Predrag-Peter Ilich
Dana College
2848 College Drive
Blair NE 68008
USA
peter.ilich.07@gmail.com

ISBN 978-3-642-04326-0   e-ISBN 978-3-642-04327-7
DOI 10.1007/978-3-642-04327-7
Springer Heidelberg Dordrecht London New York

Library of Congress Control Number: 2010921809

Cover design: KünkelLopka, Heidelberg

Printed on acid-free paper

Springer is part of Springer Science+Business Media (www.springer.com)

Preface

*The latest authors, like the most ancient,
strove to subordinate the phenomena
of nature to the laws of mathematics*

Isaac Newton, 1647–1727

The approach quoted above has been adopted and practiced by many teachers
of chemistry. Today, physical chemistry textbooks are written for science and
engineering majors who possess an *interest in* and *aptitude for mathematics*. No
knowledge of chemistry or biology (not to mention poetry) is required. To me
this sounds like a well-defined prescription for limiting the readership to a *few and
carefully selected*.

I think the importance of physical chemistry goes beyond this precept. The subject should benefit both the science and engineering majors and those of us who
dare to ask questions about the world around us. Numerical mathematics, or a way
of thinking in mathematical formulas and numbers – which we all practice, when
paying in cash or doing our tax forms – is important but should not be used to
subordinate the infinitely rich world of physical chemistry.

With this in mind, I present you a collection of problems and questions in physical chemistry, along with detailed solutions, answers, and explanations of basic
ideas. Given my personal interest I tried to focus the book on the puzzles from
the living world. My first goal is to guide you through a solution of a problem. If,
after some practice, you start understanding conventions, assumptions, and approximations that make physical chemistry, I will have accomplished my goal. Perhaps
you will then start understanding that a physical chemist is *not* someone who "uses
mathematics to subordinate the phenomena of nature" but rather an artist skilled
at finding and using assumptions and approximations that lead toward a solution
of a physical chemical problem. If you also pick up some relations, laws, rules, and
a couple of mathematical tricks along the way I will be tempted to cry – "Now I
have created a monster – s/he has become a physical chemist!" Or, perhaps I should

feel good at that point, knowing that your new knowledge and skills will point to you many doors and paths you never knew existed. It will also help you sail easier through many jobs, training programs, career changes, and other tasks you will be engaged in through your life. Read this book and study physical chemistry – for it is one of the best investments you can make in your life.

Omaha, NE, USA   Predrag-Peter Ilich

Acknowledgment

Several people, in different ways and often unknowingly, have inspired me to start this book and bring it to completion.

I would like to acknowledge my first teacher of physical chemistry: Dr. Tibor Škerlak, Professor of Chemistry at the University of Sarajevo, Sarajevo, Bosnia & Herzegovina, Yugoslavia. Tibor was a person of broad knowledge and consummate intelligence who brought physical chemistry to a high, yet unsurpassed level at the University of Sarajevo. Tragically, he fell victim to a sniper bullet, not far from the Chemistry Department, during the 1991–1995 civil war that brought the dissolution of former Yugoslavia.

During my graduate studies, Dr. Oktay Sinanoğlu, Professor of Chemistry at Yale University, and Dr. Nenad Trinajstić, Professor of Chemistry at the University of Zagreb, introduced me to the intriguing world of chemical graphs and graph-spectral tools. Dr. Ante Graovac and Dr. Tomislav Živković, from the "R. Bošković" Institute, Zagreb, Croatia, greatly helped me expand my knowledge and interest in mathematical physics and chemistry. Dr. Josef Michl, Professor of Chemistry at the University of Utah, and Dr. Patrik Callis, Professor of Chemistry at Montana State University, taught me optical and vibrational molecular spectroscopies. Years later, Dr. Maurice Kreevoy, Professor of Chemistry, University of Minnesota at Minneapolis, encouraged my interest in hydrogen bonding, while Dr. Russ Hille, Professor of Biochemistry at the Ohio State University, introduced me to the fascinating world of bioinorganic chemistry. Andonia Giannakouros, a young and very talented visual artist from Memphis, Tennessee, greatly helped me with the overall design and organization of the book. My dear friend, Dr. Nenad Juranić, Professor of Chemistry at Mayo Graduate School, taught me NMR spectroscopy and also made a number of contributions and corrections to this book. In this vein, I would like to thank the reviewers for rescuing me from some of the most egregious missteps. The errors that – alas – still persist are all but mine. I would like to thank the illustrator KünkelLopka, Heidelberg.

My students of physical chemistry provided me the joy of teaching my favorite subject and also inspired me to pursue this project. I remember them all but would like to mention two: Ms. Rosemarie Ruža Ilić Radić (no relation) who, among some 28,000 students, won the University of Sarajevo 1st prize for an original paper on dipole moments and Ms. Laurie Krusko, Loras College, Dubuque, Iowa, who valiantly and with a yet unparalleled success fought through some of the problems in this book.

Most of all, I would like to thank my beloved daughters Una and Vilma, my continuous inspiration and the joyful and challenging players in the truly delightful and ever-evolving game of the father–daughter relation.

Finally, with all my love I dedicate this little book to Eleni (who has yet to take a physical chemistry course with me).

Omaha, NE, USA   Predrag-Peter Ilich

Table of Contents

Preface v
Acknowledgment vii

Part I   Mechanics

1   Mechanical Work 3
     References 7

2   Mechanics of Gases 9
     Reference 18

Part II   Basic Thermodynamics

3   Heat Transfer 23
     References 32

4   Thermodynamics 33
     4.1   Entropy 34
     4.2   Gibbs Free Energy 40
     Reference 47

Part III   Mixtures and Chemical Thermodynamics

5   Mixtures and Solutions 53
     References 59

6   Chemical Reactions and Gibbs Free Energy 61
     References 64

7   Gibbs Free Energy and Chemical Equilibria   65
        7.1   Receptor and Ligand Equilibria   70
        7.2   Acids and Bases   87
        References   94

Part IV   Ionic Properties and Electrochemistry

8   Ions   99
        8.1   Ion Activities   102
        Reference   109

9   Electrochemistry   111
        9.1   Biological Electrochemistry   118
        References   126

Part V   Kinetics

10   Kinetics   131
        10.1   Enzyme Kinetics   140
        10.2   Reaction Barriers   145
        References   149

Part VI   Structure of Matter: Molecular Spectroscopy

11   The Structure of Matter   155
        11.1   Simple Quantum Mechanics   155
        References   169

12   Interaction of Light and Matter   171
        12.1   UV and Visible Spectroscopy   171
                12.1.1   UV/Vis Spectrophotometry   172
        12.2   Vibrational Spectroscopy   176
                12.2.1   Isotopic Effects in Molecular Vibrations   180
        12.3   Nuclear Magnetism and NMR Spectroscopy   184
        12.4   Level Population   189
        12.5   Down-Conversion of Photon Energy   192
        References   201

Index   205

Part I

Mechanics

1 **Mechanical Work** 1

2 **Mechanics of Gases** 9

1 Mechanical Work

The *physical* in physical chemistry is an important word so we start with simple physical questions and problems about lifting, pushing, twisting, or spinning. All these activities make up the topic known as *mechanical work*.

| Problem 1.1 | How strong is a sloth? |

After a nap, a teenage sloth hooked onto a tree branch in the Amazonian jungle stretches and lifts itself up to check on the neighborhood. If the sloth weighs 3.9 kg and it lifts itself 25.0 cm how much work does this take?

» Solution – Strategy

What is this question about? It is about *work*, mechanical work, similar to the work we do when helping a friend carry a piano three floors up, or the work a sloth does when it lifts itself against the force that is pulling it down. The stronger the force or the longer the distance, the more the work to be done, right? The *force* here is the gravitational acceleration or *gravity* that is pulling the sloth's body down:

$$\text{Force} = \text{body mass} \times \text{gravity} \tag{1-1}$$

Since gravity, g_n, is nature's constant and we cannot do anything about it, this leaves the body mass as the variable: the larger the mass the stronger the force. We say force is proportional to mass. Now you put everything together and punch a few keys on your calculator.

» Calculation

First, we use Newton's second law to calculate force F:

$$F = m\,(\text{mass}) \times g_n(\text{gravitational acceleration})$$

P.-P. Ilich, *Selected Problems in Physical Chemistry*,
DOI 10.1007/978-3-642-04327-7_1, © Springer-Verlag Berlin Heidelberg 2010

Body mass is 3.9 kg and gravity is given as $g_n = 9.80665$ m s^{-2} [1, 2]. You multiply these two and get the answer as

$$F = 3.9 \, \text{kg} \times 9.8 \, \text{m s}^{-2} = 38.2 \, \text{kg m s}^{-2} = 38 \, \text{N (newton)}$$

This is the force, now for the work

$$w \, (\text{work}) = \text{force} \times \text{distance} \tag{1-2}$$

$$w = 38 \, \text{N} \times 0.25 \, \text{m} = 9.56 \, \text{N} \times \text{m} = 9.6 \, \text{J (joule)}$$

So every time the sloth pulls itself up by a quarter of a meter it puts in almost ten joules of work. Question answered.

A comment: All things and beings engage in mechanical work; they don't have to have a shape – think of running water – or even be visible to the eye – think of the wind blowing, or a pesky *Bacillus streptococcus* roaming through the mucus in your throat. Even very small "things" perform mechanical work, like deep inside your biceps where the molecules that make up the muscle slide along each other when you stretch the muscle. So let us try this little riddle:

Problem 1.2 | Now cometh a little molecule.

A kinesin molecule [3] moves up with a force of 1.90 pN (piconewton = 10^{-12} N) over a distance of 25.0 nm (nanometer = 10^{-9} m). (A) Calculate the work performed by kinesin. (B) Assuming a mass of 390 kDa compare the force a sloth uses to lift itself (Problem 1.1) and a kinesin molecule uses to propel itself; which one is larger?

» **Solution A – Strategy and Calculation**

Let us do this together, step by step. The first part (A) is about work; the force is already given and we only have to pay attention to the units: pN is piconewton where *pico* stands for one part per trillion, or 10^{-12}. Likewise, the distance is given in very small units, nanometers, where *nano* stands for one part per billion, 10^{-9}. All it takes is one multiplication:

$$w = 1.9 \times 10^{-12} \, \text{N} \times 25.0 \times 10^{-9} \, \text{m} = 47.5 \times 10^{-21} \, \text{N} \times \text{m} = 4.8 \times 10^{-20} \, \text{J}$$

So the work done by this little molecule is slightly less than five times ten to the power minus twenty joules!

» **Solution B – Strategy and Calculation**

Now for the comparison – this requires a little thinking. Let us compare the two forces, the one used by the sloth and the one used by the kinesin, by calculating their ratio; we will label this ratio as $F_{1,2}$:

$$F_{1,2} = 38\,\text{N}/1.90 \times 10^{-12} = 2.0 \times 10^{13}$$

Note that there should be no unit when you divide two physical quantities. So the sloth is many, many times stronger. Yes, but this is like when you compare how much farther, than an average grasshopper, you can jump (at least we hope you can). But – think of how much bigger your body is! It's the same here – we should *compare the masses* of the sloth and the kinesin molecule to get a better idea about who is stronger.

We know the sloth's mass, let us call it "mass one," $m_1 = 3.9$ kg. The kinesin molecule's mass is "390 kDa." The "kDa" stands for "kilodalton," where each Da or *dalton* – or unified atomic mass unit, amu [1, 2] – equals $1.660538782 \times 10^{-27}$ kg. Dalton is the unit name biochemists like to use. Insert these numbers and you get the kinesin mass "m_2":

$$m_2 = 390 \times 10^3 \times 1.66 \times 10^{-27}\,[\text{kg}] = 6.47 \times 10^{-22}\,\text{kg}$$

Now compare the two masses, m_1 (sloth) and m_2 (kinesin), the same way you compared the two forces; we will label this ratio $m_{1,2}$:

$$m_{1,2} = 3.9\,[\text{kg}]/6.5 \times 10^{-22}\,[\text{kg}] = 6.0 \times 10^{21}$$

The sloth has a 6.0×10^{21} larger body mass than kinesin. If the sloth is as strong as the kinesin molecule it should be about 6.0×10^{21} times stronger. But it is not, it is only 2.0×10^{13} times stronger and if we compare the *ratio* of the forces with the *ratio* of the body masses we will get a *ratio or ratios*:

$$F_{1,2}/m_{1,2} = 2.0 \times 10^{13}/6.0 \times 10^{21} = 3.3 \times 10^{-9}$$

So the teenage sloth is only 3-billionth as strong as (an adult) kinesin molecule. A wimp, wouldn't you say?

A note on work and path: You are helping your friend move to a new apartment on the third floor. You grab a chair from the moving truck and carry it upstairs. Let us label the work you have done as $w(1)$. Then you grab another chair and carry it the same way; the work is $w(2)$. But these are not really heavy chairs and you could take both chairs at once and carry them upstairs; the work in this case will be $w(1 + 2)$. Now if you think of the chairs alone, then the sum of $w(1)$ and $w(2)$ will be the same as the work $w(1+2)$. So we can say that work is additive. It turns out that in most cases this is not so. Consider this: you grab the chair and carry it upstairs. Not knowing where your friend's apartment is you go all the way to the fifth floor and then, realizing your mistake, climb down and deposit the chair on the third floor. What is the work transferred to the chair? It is $w(1)$. And what is the work you have done? It is a lot more than $w(1)$; it is also different than in the first case when you went straight to the third floor. We are going to state this in English:

1

In general, work depends on the path along which it is carried out and on the destination point.

We say that work is path dependent or that work is a path function. Make a note: *path function*. From experience you know there are more and less difficult ways to lift an object, depending on how you use your body to carry out this work. The following grocery shopping scene illustrates this.

Problem 1.3	Whole milk only.

Done with shopping, you are in a parking lot transferring the merchandise to your car. You grab a large jug of milk, lift it, and put it in the car trunk. Now try lifting it with your arm, without bending the elbow. Do you make the same effort in this case?

» **Solution – Strategy and Calculation**

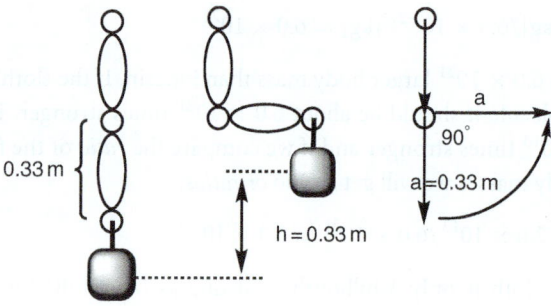

■ **Fig. 1.1** The geometry of the forearm lever

Look at Fig. 1.1: When I bend my elbow I use my forearm, a 33 cm lever, to lift the jug of milk. If I think of my elbow as an axis of rotation I can say that I have rotated my forearm by 90°. The rotational component of the work I have performed is given by *torque*. It is given as

$$torque = F \times \text{lever length} \times \text{sinus (angle of rotation)} \qquad (1\text{-}3)$$

The expression takes a particularly simple form in the case of the 90° angle:

$$\tau = F \times r \times \sin 90° = F \times r \times 1$$

Let us calculate how much work I have done. Let us assume that the mass of a large jug of milk is 4.00 kg. Then the torque needed to lift the milk jug by bending my elbow will be

$$\tau(\text{elbow}) = 4.00\,[\text{kg}] \times 9.81\,[\text{m s}^{-2}] \times 0.33\,[\text{m}] = 39.2\,[\text{kg m s}^{-2}]$$
$$\times 0.33\,[\text{m}] = 12.9\,\text{N m}$$

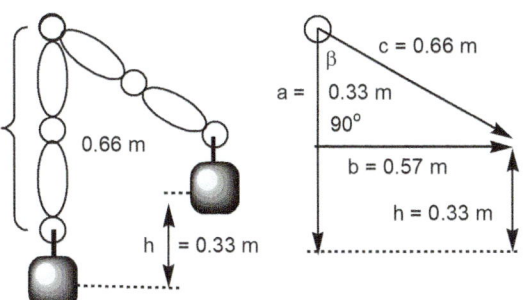

■ **Fig. 1.2** The geometry of using the whole arm as the lever

And now, just to show how strong I am, I use the whole arm to lift the same jug (Fig. 1.2). The work received by the jug is the same: it changed the height by 0.33 m upward. But what about torque?

The formula for torque is the same:

$$\tau = F \, (\text{force}) \times r \, (\text{lever}) \times \sin \, (\text{rotation angle})$$

The geometry is similar too: I use the whole arm length, lever = 0.66 m, to lift the same jug by 0.33 m. The only thing I do not know is the angle of rotation, β. I will find it using the Pythagorean rule for b:

$$b = (c^2 - a^2)^{1/2} = 0.57 \, \text{m}$$

And the *sine rule* for triangles tells me this:

$$\sin \beta = 0.57 \, [\text{m}] \, (\sin 90°/0.66 \, [\text{m}]) = 0.86$$

When I insert these numbers into the formula for torque, I get

$$\tau \, (\text{shoulder}) = 4.00 \, [\text{kg}] \times 9.81 \, [\text{m s}^{-2}] \times 0.57 \, [\text{m}] \times 0.86 = 19.2 \, \text{N m}$$

This is 19.2/12.9 = 1.50 or 50% more work than when I lifted the milk jug using elbow only. But, then, we have all known this – you never use a whole arm's length to lift something. Only now you can show it by numbers and this is an important goal of physical chemistry: learn to put numbers behind words like *less*, *more*, or *a lot*.

References

Units and Constants
1. Woan G (2003) The Cambridge handbook of physics formulas. Cambridge University Press, Cambridge
2. National Institute of Standards. URL: http://physics.nist.gov/constants , accessed July 31, 2009

Mechanics of Molecules
3. Kuo SC, Sheetz MP (1993) Force of single kinesin molecule measured with optical tweezers. Science, 260:232–234

2 Mechanics of Gases

A note on the states of matter: Physical chemists are very concerned with *states of matter*: gases, liquids, solids, etc.; are there more states there? Yes – think of the chicken soup (suspension) you are warming up on the gas flame (plasma) while checking your watch display (liquid crystal), or think of the peanut butter (emulsion)–jam (gel) sandwich you had this morning. You may think of the shaving cream (foam) or the cologne spray (aerosol) in your bathroom. You may think of what the inner parts of your body are made of or – why limit our horizon to little things only? – what did they say black holes are made of? As a matter of fact most of the world around us – including ourselves – is built of one of these "other" states of matter. Historically, physical chemistry has been developed with the three "pure" states – gas, liquid, solid – in mind and has only recently started making inroads into the intriguing and very complex world of the other states of matter.

Gases are simplest and have been studied most. You may think of your car: a flat tire (no air), the (noxious) exhaust gases, the anti-collision air bag (in the movies only, please). You may think of a balloon: a birthday balloon, a hot air balloon, or perhaps you just read about a meteorological balloon. Or you may think of what you do all the time: inhaling and exhaling, about ten times a minute (or longer – if you are waiting to exhale). We will do short examples of each.

Problem 2.1	Mr. Bond's latest assignment (in Kazakhstan).

On a cold, $t = -3°F$, late winter day in Kazakhstan, Cyril Bond checks the pressure in the front left tire on his Aston Martin–Hyundai Accent (special edition) service car and finds that it reads only 17 pounds per square inch, psi. After consulting with Nurudin, a garage attendant, Cyril asks him to add more air, until the pressure gauge reads 2.3 kg/cm^2, or "atmosphere," atm (Nurudin is using a metric pressure gauge).

Question (A): What was the pressure in Mr. Bond's car tire, in psi, after he left the garage?

P.-P. Ilich, *Selected Problems in Physical Chemistry*,
DOI 10.1007/978-3-642-04327-7_2, © Springer-Verlag Berlin Heidelberg 2010

2

Question (B): What will the pressure be in the tire at 118°F, when Mr. Bond hits the streets of Calcutta, India, on his next assignment, 6 weeks later? How much higher will this pressure be than the recommended, 30 psi? Carefully list all assumptions made. Show the unit conversions clearly.

» Solution A – Strategy

This is a relatively simple problem made complicated through use of different units. Although most of the world today uses metric system of units the mass, size, and volume of certain objects used or mentioned everyday are expressed in non-metric, historically based common units. We still hear or read about the price of a barrel of crude oil; the repairman we hired needs a two-by-four inches board, and we go to a grocery store to buy a dozen eggs, not to mention time, which has resisted all attempts of conversion to the base-10 system. Sciences have not been spared this diversity; we are still consuming (and burning) calories, measuring very small distances in Ångstroms, or comparing atomic energies in electronvolts. The use of different units is an annoying feature in physical chemistry but it is something you have to learn to live with; it is perhaps a little bit like foreign words you learn to use along with your native tongue.

Let us remind ourselves what is pressure: a force, F, applied at a certain spot or area, A:

$$p = F/A \tag{2-1}$$

The larger the force, F, the larger the pressure. We say that pressure is *directly proportional* to the applied force. On the other hand, the smaller the area, the smaller the number that divides the force, and pressure will be larger. We say that the pressure and area are *inversely proportional*.

A note on intensive and extensive properties: Many things we are surrounded with can be added or subtracted: money, apples, liters of gas. Mass, distance, volume, and time can accumulate and this is how we label them: *cumulative properties*. We can also use the words extrinsic or extensive for these properties. Pressure, on the other hand, depends on two quantities, two properties: force and area. We say that pressure is a *composite* property. For this reason we cannot add, subtract, or multiply two pressures. There are other properties that cannot be added, subtracted, and multiplied with each other and we have a common name for them: *intensive properties*. Other words with the same meaning as intensive are intrinsic or specific.

Make a note: *cumulative* and *intensive* properties.

» Calculation A

In the garage in Kazakhstan, $p = 2.3$ atm, or $p = 2.3$ kg/cm^2. Convert kilograms to pounds and square centimeters to square inches and you will have the first answer:

$$p = (2.3\,[\text{kg}] \times 2.2\,[\text{lb/kg}])/(1\,[\text{cm}^2] \times 0.155\,[\text{inch}^2/\text{cm}^2]) = 32.6 \approx 33\,\text{lb/inch}^2$$

The answer to the question (A) is 33 psi, within about 1% error.

» Solution B – Strategy

Now this is a trickier problem and you will have to consult a physical chemistry textbook, under the chapter on "Ideal Gases." There is a good and useful formula for gases, most of which behave like *ideal gases*:

Pressure × volume = # moles × gas constant × temperature

When translated into physical chemical symbols, this expression reads

$$pV = nRT \qquad\qquad\qquad (2\text{-}2)$$

You can use this formula to calculate many properties of gases and find answers to many questions in physical chemistry, as well as in everyday life. You should first figure out which of the properties in the gas equation changes and which remains constant. "R" – is the so-called gas constant; nothing to change there, you only have to be careful about the units you use for R. "n" – is the amount of the gas, given in *moles*, which you have to figure out from the liters, kilograms, or ounces of gas. (By the way, converting all amounts to moles is a very good way to go about chemical calculations.) Unless there is a chemical reaction involved the number of moles usually does not change through a problem. This leaves p, V, and T and they do depend on each other and are subject to change. For one temperature we will have one pressure and one volume, for a different temperature different pressure and volume. So you may write

At $-3°$F (winter in Kazakhstan): $p_1 V_1 = n_1 R T_1$
At $+118°$F (spring in Calcutta): $p_2 V_2 = n_2 R T_2$

Now we have to make a couple of *assumptions*:

- First, we assume that the tire does not leak air.
- Second, we assume that volume of the air in the tire does not change.

The first assumption is quite sound: if the tire is in good condition it should not leak for weeks and months. This is telling us that the amount (i.e., the number of moles) of air in the tire during the winter days in Kazakhstan, n_1, is the same as the amount

2

of air during a hot spring day in Calcutta, n_2. When you translate this statement into a formula you will write $n_1 = n_2$, or just n, the number of moles.

The second assumption, that the volume of the tire is the same at $-3°F$ and at $+115°F$, is a little less sound. If you want better information you should consult an expert or a reliable source on car tires. I would say – mainly from experience – that the volume changes a little but not so much that we should worry about it in this problem. We may check this issue later but for the moment let us assume that $V_1 \approx V_2$, or just V, the volume.

» **Calculation B**

Let us re-write the previous two equations using n and V:

$$p_1 V = nRT_1 \tag{A}$$

$$p_2 V = nRT_2 \tag{B}$$

Now you may go back and read the second part of the problem again. The question is What is the pressure going to be in the tire while in Calcutta, at $118°F$? Another way to ask the question is What is p_2 going to be? Let us try to solve the problem. You should make two lists: (a) a list of the things you know (or can find out) and (b) a list of the things you do not know but have to figure out.

(a) p_1, T_1, and T_2
(b) p_2

I suggest you also make a third list: (c) a list of things you don't need to know. This may turn out to be an important list.

(c) n, V, and R

The value of R, the gas constant, you can find easily. This leaves us with n, the number of moles of gas in the tire, and V, the volume of the air in the tire. Nobody is really asking you about either n or V and you should try to get them out of the way.

You can do this by applying *little big trick* #1: divide two equations and cancel the same terms. You should divide the left side of (A) with the left side of (B) and then divide the right side of (A) with the right side of (B):

$$p_1 \times V / p_2 \times V = n \times R \times T_1 / n \times R \times T_2$$

Now cancel the same quantities above and below the dividing line; this will leave you with a simplified equation:

$$p_1/p_2 = T_1/T_2$$

You need p_2 so you will re-arrange the equation (that is, first turn it upside down and then move p_1 to the other side of the $=$ sign):

$$p_2 = p_1 \times (T_2/T_1)$$

Insert the numbers for p_1, T_1, and T_2. But – be careful! You will have to convert the temperatures given in degrees Fahrenheit to the physical chemical temperature scale, given in degrees Kelvin, K. Consult a textbook and you will find the following conversion: $T\,[\mathrm{K}] = (t\,[\mathrm{F}] - 32/1.8) + 273.2$. Pretty complicated, isn't it? So punch few keys on your calculator and you will get for the temperature in Kazakhstan, T_1, and Calcutta, T_2:

$$T_1 = (-3 - 32)/1.8 + 273.2 = 253.7\,\mathrm{K}$$

$$T_2 = (118 - 32)/1.8 + 273.2 = 320.9\,\mathrm{K}$$

The pressure p_2 now reads

$$p_2 = 2.3\,\mathrm{kg\,cm}^{-2} \times 320.9\,\mathrm{K}/253.7\,\mathrm{K} = 2.9\,\mathrm{kg\,cm}^{-2}$$

Question (B) was the following: what is the pressure reading in psi? You will have to convert $2.9\ \mathrm{kg\ m}^{-2}$ to pounds per square inch in the same way you did it in part (A) of this riddle:

$$p_2[\mathrm{psi}] = (2.9\,\mathrm{kg} \times 2.2\,[\mathrm{lb/kg}])/(\mathrm{cm}^2 \times 0.155\,[\mathrm{inch}^2/\mathrm{cm}^2]) = 41.3\,[\mathrm{psi}]$$

And how much higher this is than 30 psi — the recommended pressure in car tires? This is straightforward: subtract 30 from 41.3 and you get 11.3 psi. A whole eleven point three pounds per square inch! This is a seriously overinflated tire. Careful, Mr. Bond!

A comment: I suggest we take a little break now – this was a lot of work. Afterward, we can look at other problems involving gases and a question about hot air ballooning which is both entertaining and useful. Hot air balloons do look beautiful and if you have taken a ride in one – and come down safely – you know what I mean.

| Problem 2.2 | Three men in a balloon (to say nothing of the dog). |

It is a late fall, slightly chilly morning, $t = 10.0°\mathrm{C}$. Three men and a dog drive a couple of miles until they reach an empty farmland area where a large balloon is anchored. They climb into a sturdy wicker gondola equipped with propane cylinders and a twin gas burner, all attached to a large spherical balloon, and decide to take a flight. They slide their aviator goggles and put leather gloves on, ignite the burners, and get ready for a liftoff (no goggles for the dog). The passengers, dog, basket, ropes, burners, gas cylinders, and the nylon–Nomex® envelope of the balloon weigh 678 kg. (Typical weight of a mid-size balloon, a basket with –three to five passengers, a twin burner, and –two to four gas cylinders is 650–750 kg [1].)

2

As the air inside the balloon becomes warmer it becomes less dense and when its buoyancy exceeds the net weight of the balloon and the cargo, the balloon will take off. Given that the fully inflated balloon is a nearly perfect sphere of 9.0 m radius what should be the average temperature of the air inside the balloon [°C] needed for a liftoff? Keep in mind that the balloon is open and the pressure of the air inside the balloon equals the pressure of the air outside the balloon.

» Solution – Strategy

The hot air balloons are based on the principle of buoyancy. For example, a wooden spoon set free in the air falls down because its buoyancy is smaller than its mass (weight). In water, it is the opposite case: the buoyancy is larger and the spoon floats. In hot air balloons the medium is always the same – air. We only change its density by changing its temperature. When the density of the air inside the balloon equals the density of the air outside, its buoyancy is zero and its net mass – 678 kg in this case – will prevent it from taking off. We have to make the air inside the balloon less dense by heating it up, e.g., using propane burners.

» Solution – Calculation

Assume that the balloon is filled with the air at 10°C and calculate the volume of the balloon. The balloon is assumed to be spherical so find the formula for volume of a sphere and insert the value for radius, $r = 9.0$ [m]:

$$V = 4 \times \pi \times 9.0^3 \ [\text{m}^3]/3 = 3,054.6 \ \text{m}^3 = 3.055 \times 10^3 \ \text{m}^3$$

This is the *volume* of the balloon – three thousand and fifty cubic meters. What you need to know is how many *moles* of air there are in 3,055 m³. You will get this by reshuffling the $pV = nRT$ equation in the following way: $n = RT/pV$. But first you have to convert the temperature to the thermodynamic scale, [K], by adding 273.2 to the temperature in degrees Celsius (or centigrades): $T = 10.0 + 273.2 = 283.2$ K. Now you can calculate n:

$$n_1 = 101{,}325 \ [\text{N m}^{-2}] \times 3.055 \times 10^3 \ [\text{m}^3]/8.314 \ [\text{J K}^{-1} \text{mol}^{-1}]$$
$$\times 283.2 \ [\text{K}] = 1.315 \times 10^5 \ \text{mol}$$

Multiply the number of moles of air by its molar mass and you will have the mass of all air inside the balloon:

$$m_1 \ (\text{cold air}) = 1.315 \times 10^5 \ [\text{mol}] \times 0.029 \ \text{kg mol}^{-1} = 3{,}813.5 = 3{,}814 \ \text{kg}$$

We used the following value for the molar mass of air: $m_m(\text{air}) = 28.97$ g mol^{-1} = 0.029 kg mol^{-1}. This is an *average* molar mass for the mixture of 79% of N_2, 20% of O_2, 1% of Ar, and a wisp of other gases.

What should we do now?

The mass of the cold air, $T_1 = 10°C = 283.2$ K, inside the balloon is 3,814 kg. Since the balloon is surrounded by the air of same temperature and density, the *buoyancy* of the air inside the balloon is *zero* and, given only the air, the balloon could move left or right, but not lift. But what keeps it grounded is the 678 kg of the cargo mass. The balloon, with the basket and passengers, will be able to start lifting off when the buoyancy of the air inside the balloon equals the cargo mass, $m_0 = 678$ kg. Let us call this the second mass, m_2:

$$m_2 = 3,814 - 678 = 3,136 \text{ kg}$$

How do you make the air inside the balloon weigh 3,136 kg instead of 3,814 kg? Simple – you heat it up! When the temperature of a gas increases, its density, therefore its total mass – *decreases* (assuming the pressure stays the same; this is also known as Charles' law). So the question you have to answer is, At what temperature the density of the air inside the balloon will decrease so much that its mass is 3,136 kg or less. You will use the $pV = nRT$ equation to find this temperature, let us call it T_2. But as you can tell there is no place for the mass of air in the gas equation; you need the number of moles. Let us call it n_2:

$$n_2 = 3,136 \text{ kg}/0.029 \text{ kg mol}^{-1} = 1.081 \times 10^5 \text{ mol}$$

So n_2 is the number of moles of the hot air inside the balloon; you may insert this number in the gas equation. You also know the pressure and volume of the hot air balloon – it is unchanged: p_2 at temperature T_2 is the same as p_1 at temperature T_1. The same for volume; we just drop the indices and use p and V. Now you know everything you need to know to calculate the temperature T_2:

$$T_2 = pV/n_2R$$

Insert the number of moles, n_2, and calculate the temperature, T_2:

$$T_2 = 101,325 \, [\text{N m}^{-2}] \times 3.055 \times 10^3 \, [\text{m}^3]/1.081 \times 10^5 \, [\text{mol}]$$
$$\times 8.314 \, [\text{J K}^{-1} \text{mol}^{-1}] = 344.4 \text{ K}$$

When you convert this back to degrees Celsius you will see the air inside the balloon gets fairly warm:

$$T_2 = (344.4 - 273.2) = 71.2°C$$

Now that was quite a workout, wasn't it? Let us take a break before we go on to the next question.

2

Problem 2.3 | Hot air ballooning, a sequel.

Understanding how hot air balloons float helps you understand how a human body, with lungs filled with air, floats in water. I suggest you practice a little more by changing the conditions in the previous riddle and solving it by yourself. Let us assume it is a spring day with the air temperature (outside and inside the balloon) 20.0°C and the same group of ballooners getting ready to take off. They want a speedy takeoff and are going to make the air inside the balloon have the buoyancy equal to the cargo mass, 678 kg, plus another 10%. (So m_2 will be equal to $m_1 - 678 \times 1.1$ kg.) How hot will the air inside the balloon have to be?

Answer: $\mathbf{T}_2 \approx 87°C$

Problem 2.4 | Waiting to exhale.

The partial pressure of oxygen in the inhaled air, pO_2 (in) $= 159$ mmHg, and in the exhaled air, pO_2 (ex) $= 116$ mmHg. Assuming that the air pressure, p (air) is 760 mmHg, calculate how many grams of O_2 are transferred from the atmosphere to our alveoli each minute of normal breathing (10 inhalations at 2.0 L each).

» **Solution – Strategy**

We will of course start with the $pV = nRT$ equation and use it to calculate the number of moles of O_2 during the inhalation and then the number of moles of exhaled oxygen, subtract the two values and the difference will give us the answer. However, there are more "things" in this problem and we will have to solve them one by one. Let us first calculate the number of inhaled and exhaled moles of O_2.

» **Solution – Calculation**

Let us write the gas equation for inhalation and express it for n, the number of moles:

$$p_1 V_1 = n_1 R T_1 \qquad n_1 = p_1 V_1 / R T_1$$

You may now insert the numbers for p_1, V_1, and T_1, find the value for R, and calculate n_1. Or you may use a little shortcut. Let me explain what I have in mind. At room temperature and the pressure of 760 mmHg (which equals 101,325 Pa) a mole of gas, any *ideal-like* gas, has a volume of 24.79 L; in physical chemistry textbooks this volume, derived from the so-called Avogadro's law, is known as the volume of one mole of gas at normal temperature and pressure, NTP (which is slightly different from the standard temperature and pressure, STP – but we shall not bother with the little physical chemistry obfuscations). Let us call this quantity n (1 mol) and write

$$n(1\,\text{mol}) = 24.79\,\text{L mol}^{-1}$$

The riddle says that each minute we inhale (on average) ten times and each time the volume of the inhaled air is 2 L which makes $10 \times 2\,\text{L} = 20\,\text{L}$. So $V_1 = 20\,\text{L}$. Well, if one mole of gas takes 24.79 L then there is obviously less than one mole of gas in 20 L of air. How much less? Divide the smaller number by the bigger number and you will get

$$n_1 = 20.0\,\text{L}/24.79\,[\text{L mol}^{-1}] = 0.8068 = 0.807\,\text{mol}$$

Lest we forgot, these are the moles of air; what we need are the moles of O_2.

» More strategy

Now we have to look at air as a *mixture of gases*. Indeed, there is nitrogen, oxygen, argon, water vapor, and other gases and vapors in the air. Each of these is a part, a *fraction* of the total. How big a fraction? If all of the air were just O_2 then the fraction of O_2 would be 100%/100% = 1. But you know it is less than that; you are given the numbers which tell you what is the fraction of O_2 in the air. Check a textbook and find the *Dalton law* of partial pressures. It says that the total pressure is a sum of partial pressures:

$$p_{\text{tot}} = p_1 + p_2 + p_3 + \cdots \qquad (2\text{-}3)$$

In the case of inhaled and exhaled air this will be given by the following expression:

$$p_{\text{tot}} = p(O_2) + p(N_2) + p(Ar) + p(CO_2) + p(H_2O) + \cdots$$

How do we know partial pressures? The partial pressure of O_2 (an ideal-like gas), in a mixture with other ideal-like gases, is the same as the number of moles of O_2 divided by the number of moles of all gases in the mixture. We call this ratio a *molar fraction* and label it by x: $x(O_2) = n(O_2)/n_{\text{tot}}$. Another way to find molar fraction of, for example, O_2, is by dividing its partial pressure by the total pressure:

$$x(O_2) = \text{partial pressure }(O_2)/\text{total pressure of all gases in the air}$$

» More calculation

Go back to the problem and read off these numbers, $p(O_2) = 159\,\text{mmHg}$ and $p(\text{air}) = 760\,\text{mmHg}$. You can now go and convert the pressures in mmHg into Pa or, since both units are the same, divide the two numbers and cancel the units; molar fractions have no units.

$$x(O_2) = 159\,\text{mmHg}/760\,\text{mmHg} = 0.209$$

So 0.209th part or 20.9% of air is dioxygen, O_2. Given that each minute you inhale 20 L or 0.807 mol of air (as we calculated above) the number of moles of O_2 will be 0.209th part of it:

$$n(O_2,\ \text{in}) = 0.209 \times 0.807\,\text{mol} = 0.1688 = 0.169\,\text{mol}$$

What now? Do the same calculation for the exhaled air:

$$n(O_2,\ \text{ex}) = (116\,\text{mmHg}/760\,\text{mmHg}) \times 0.807\,\text{mol} = 0.153 \times 0.807\,\text{mol} = 0.123\,\text{mol}$$

So in 1 min you inhale 0.17 mol O_2 and exhale 0.12 mol O_2; clearly, the difference is what is "left" in the lungs and transferred to the bloodstream – via the protein molecules called hemoglobin, HbA – and passed to the cells in our body. (The cells, of course, use O_2 to burn the nutrients and return the CO_2 gas.) The difference in the moles of dioxygen, Δn, is given as

$$\Delta n(O_2) = 0.169 - 0.123 = 0.046\,\text{mol}$$

The number of moles, multiplied by the molecular mass of O_2, will give you the mass of dioxygen in grams:

$$m(O_2) = \Delta n[\text{mol}] \times 32.00[\text{g mol}^{-1}] = 1.472\,\text{g}$$

Not much, don't you think so?

A note on gas equations: The $pV = nRT$ is a good equation and will take you a long way with gases like helium at room or at higher temperatures and normal or lower pressures. However, with industrial gases like propane, C_3H_8, or sulfur hexafluoride, SF_6, at high pressures and low temperatures the interaction between the gas molecules becomes significant. Also, at higher pressures and lower temperatures the volume of the (very tiny but also numerous) gas molecules becomes non-negligible. These two effects – the molecule–molecule interaction and the cumulative volume of all molecules – are not accounted for in the ideal gas equation and the $pV = nRT$ relation becomes less and less adequate (e.g., negative pressures and other non-physical results are obtained). For these cases gas equations "corrected" by second- and higher-order terms are used, like the van der Waals or Dieterici equations. This, however, is more a matter of technical thermodynamics.

Reference

1. Hot air ballooning. URL: http://en.wikipedia org/wiki/Hot_air_balloon. Accessed July 31, 2009

Table I

Summary of the mechanics of solid bodies and gases – what have we learned?

Review the material we have covered and write down the definitions, equations, relations, and quantities you deem important. You may think of it as your own glossary of terms. One day you will use it to write your own textbook.

Words and Phrases	Symbols, Formulas and Numbers
Mass (kg), length (m), volume (L), time (s)	$1 \text{ kg} = 10^3 \text{ g}; 1 \text{ m} = 10^2 \text{ cm}$ $1 \text{ L} = 10^3 \text{ mL} = 10^{-3} \text{ m}^3 = 10^3 \text{ cm}^3$
Kilo (k), Mega (M), Giga (G), Terra (T)	$k = 10^3, M = 10^6, G = 10^9, T = 10^{12}$
Deci (d), centi (c), milli (m), micro (μ), nano (n), pico (p)	$d = 0.1, c = 0.01, m = 10^{-3}, \mu = 10^{-6},$ $n = 10^{-9}, p = 10^{-12}$
Force is mass *times* acceleration, unit = newton, N	$F = m \times a \ (g_n) \text{ [N]}; g_n = 9.8066 \text{ m s}^{-2}$
Work is force *times* distance (height), unit = joule, J	$w = F \times l \text{ (or } h) \text{ [J} = \text{N m]}$
Torque is force *times* lever length *times* sinus angle, τ	$\tau = F \times r \times \sin \Theta \text{ [N m]}$
Pressure is force *divided* by area, unit = pascal, Pa	$p = F/A; p_0 = 101{,}325 \text{ N m}^{-2} \text{ [Pa]}$
Gas (ideal) equation	$pV = nRT$
Volume of 1 mol of gas at standard temperature and pressure	$n \text{ (1 mol)} = 24.79 \text{ L at } T = 298.15 \text{ K}$ and $p = 101{,}325 \text{ Pa}$
Thermodynamic temperature, unit = kelvin, K; Gas constant, R	$T = \text{deg}°\text{C} + 273.15 \text{ [K]}$ $R = 8.314472 \text{ J K}^{-1} \text{ mol}^{-1}$
Total pressure as a sum of partial pressures	$p_{tot} = p_1 + p_2 + p_3 + \cdots$
Partial pressures and molar fractions	$p_1 = x_1 \times p_{tot}$ $x_1 = n_1/n_{tot}$ $x_1 + x_2 + x_3 + \cdots + x_{last} = 1.0$

Table 1

Summary of the mechanics of solid bodies and gases – what have we learned?

Review the material we have covered and write down the definitions, equations, relations, and quantities you deem important. You may think of it as your own glossary of terms. Maybe you will use it as your textbook.

Part II

Basic Thermodynamics

3   Heat Transfer 23

4   Thermodynamics 33

3 Heat Transfer

We are now stepping into the traditional physical chemistry area known as *thermodynamics*. Thermodynamics is concerned with a change of the total energy of a system. A *system* is a body which performs work ($-w$) or gives away heat ($-q$); the same body can also receive work ($+w$) or heat ($+q$). Notice the sign: when you carry out work or give away heat (you hold someone's cold hands) the sign is minus. When you receive work (a friend carries your schoolbooks) or heat (warming up your hands on a campfire) the sign is plus. Come to think – it is the same with personal finances: the money given definitely has a minus sign. In the next few examples we will look into what happens when *heat* is exchanged between two bodies.

Problem 3.1 | A mountain saga.

A mountain climber runs into a sudden rainstorm, t(rain) $= 6°C$, and, unable to find a shelter, she becomes completely soaked with cold rain. The climber weighs 65.0 kg and her clothes have absorbed 1.20 kg of rainwater. Assume that the heat capacity of her body is equivalent to the heat capacity of water and calculate how much heat she has lost in this event.

≫ Solution – Strategy

Let us illustrate this situation by a simple scheme: a box which represents the climber, $m_1 = 65$ kg, and the clothes around the body, getting soaked by cold rain, $m_2 = 1.2$ kg, Fig. 3.1. When you plot the absolute temperatures of the climber, $T_1 = 36.6°C + 273.2 = 309.8$ K, and the rain-soaked clothes, $T_2 = 6°C + 273.2 = 279.2$ K, you get a diagram as in Fig. 3.2.

Once her clothes are soaked the climber and the rain are in contact and the heat flows from the climber to the rain-soaked clothes, Fig. 3.3. The climber *loses* heat and becomes colder while the rain-soaked clothes *gain* heat and become warmer. The mountain climber, at $T_1 = 309.8$ K before the rain, will lose heat and drop the

P.-P. Ilich, *Selected Problems in Physical Chemistry*,
DOI 10.1007/978-3-642-04327-7_3, © Springer-Verlag Berlin Heidelberg 2010

3

WET CLOTHES: 1.2 kg

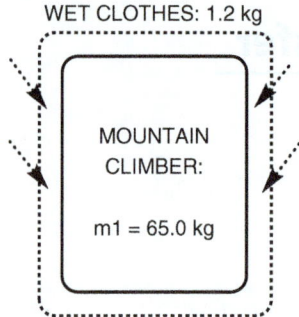

MOUNTAIN
CLIMBER:

m1 = 65.0 kg

◻ **Fig. 3.1** The masses of the two systems

T1(climber) = 309.8 K

ΔT= 30.6 K

T2(rain) = 279.2 K

◻ **Fig. 3.2** The temperatures in [K] of the two systems

RAIN SOAKED CLOTHES

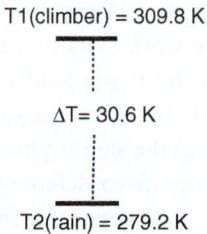

MOUNTAIN
CLIMBER

HEAT
flow:

◻ **Fig. 3.3** The direction of the heat flow

temperature to T_f, the *final temperature*. The rain in the clothes, initially at $T_2 =$ 279.2 K, will get warmer and reach the same final temperature, T_f. The two bodies – the climber and the rain-soaked clothes – are now in *thermal equilibrium*, Fig. 3.4.

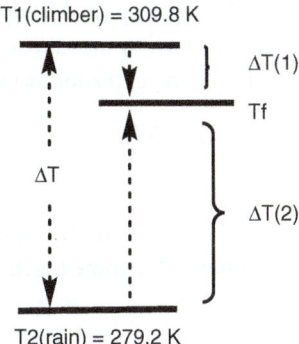

T1(climber) = 309.8 K

ΔT(1)

Tf

ΔT

ΔT(2)

T2(rain) = 279.2 K

◼ Fig. 3.4 Initial and final temperatures

» Solution – Calculation

We start by repeating the following statement about heat transfer: the heat q_1 flows away from the body. The heat passed to the rain-soaked clothes is q_2. Now we make a very simple statement: q_1 and q_2 are the same and write

$q_1 = q_2$

This does not look like a very deep statement but it actually is – and it helps us solve the problem. Our experience and physical measurements tell us that heat – the amount of heat in a body – is determined by the following three quantities: (1) the mass, m (the bigger the body the more the heat), (2) the property of the material called *heat capacity*, C_p, and (3) the difference in the temperature before and after the heat transfer, ΔT. Translated into physical chemical symbols this sentence now reads as

$q = m \times C_p \times \Delta T$ (3-1)

The mass of a body can be easily measured – and seen – and the temperature could be measured by thermometer or felt by touch. But heat capacity is not something obvious, not something intuitive; by looking at or touching two different things – a glass of milk and a slice of bread – you have no ways of telling which of the two has a higher heat capacity. We have to accept it as a property of matter – like density, like color – and use the experimentally determined and tabulated heat capacity values in all problems involving heat transfer. So for q_1 and q_2 you can write

$q_1 = m_1 \times C_p(1) \times \Delta T_1$

$q_2 = m_2 \times C_p(2) \times \Delta T_2$

Remember that we just said that q_1, the heat passed from the body to the rain-soaked clothes, is the same as q_2, the heat received by the rain-soaked clothes from the body. Translate the sentence into symbols and get the following very useful expression:

$$m_1 \times C_p(1) \times \Delta T_1 = m_2 \times C_p(2) \times \Delta T_2 \tag{3-2}$$

» Assumptions
Now comes the important part of making and using assumptions, that is, where common sense, experience, and knowledge come together.

» Assumption #1
The problem states (read it again) that "the heat capacity of the human body is approximately equal to the heat capacity of water." This is an assumption and approximation but it is not so crude; most of the human body is water, after all.

» Assumption #2
The problem states that the heat capacity of the rain-soaked clothes is similar to the heat capacity of water; not true for the clothes but perfectly true for the 1.2 kg of cold rainwater. As we do not care about the clothes this is a good assumption.

These two assumptions will push you a long way toward solving the problem. What they say is that $C_p(1) = C_p(\text{water})$ and also $C_p(2) = C_p(\text{water})$. So $C_p(1)$ and $C_p(2)$ are the same! Go ahead and cancel them! The previous Equation now reads

$$m_1 \times \Delta T_1 = m_2 \times \Delta T_2 \tag{3-3}$$

We know what m_1 and m_2 are but we do not know ΔT_1 and ΔT_2; so you have *one equation* with *two unknowns*. An intractable problem, mathematicians would say.

Yes, but there is something else we know and Fig. 3.4 tells us what this is. Look at it and figure out the sum of ΔT_1 and ΔT_2. It equals ΔT, the difference between the body temperature and the rain temperature before any heat has been exchanged. From Fig. 3.2 you find this temperature difference: $\Delta T = 30.6$ K. Now, use symbols and numbers to re-write what we just said:

$$\Delta T_1 + \Delta T_2 = 30.6 \,\text{K} \tag{3-4}$$

It is time to use the powerful *little big trick #2*: when two unknowns, ΔT_1 and ΔT_2 in this case, are connected with an additional relation, (3-4) in this case, use this relation to express one of the unknowns with the help of the relation. In this way you remove the other unknown. You will use (3-4) to write

$$\Delta T_2 = 30.6 \,\text{K} - \Delta T_1$$

Now copy (3-3) and replace ΔT_2 with 30.6 K $- \Delta T_1$ from (3-4) and you get the following relation:

$$m_1 \times \Delta T_1 = m_2 \times (30.6 \,\text{K} - \Delta T_1)$$

What have you achieved using this trick? You have eliminated one unknown and now you have one equation with only one unknown quantity: ΔT_1. Multiply what has to be multiplied and move all expressions containing ΔT_1 to the left side of the equation:

$$m_1 \times \Delta T_1 - m2 \times \Delta T_1 = m_2 \times 30.6\,\text{K}$$

This gives you

$$\Delta T_1 = 30.6 \times m_2/(m_1 + m_2)$$

$$\Delta T_1 = 30.6[\text{K}] + 1.2[\text{kg}]/(65.0[\text{kg}] + 1.2[\text{kg}]) = 0.55\,\text{K}$$

So the mountain climber loses a little more than half a degree centigrade (this will make her feel a little chilly). It is easy now to calculate q_1, the heat the climber has lost:

$$Q_1 = 65.0[\text{kg}] \times C_p(\text{water}) \times 0.55[\text{K}]$$

The $C_p(\text{water})$ is found to be $4.187 \times 10^3\,\text{J K}^{-1}\,\text{kg}^{-1}$ which will give you

$$Q_1 = 150{,}960\,\text{J} = 151\,\text{kJ}$$

Done. That was quite a riddle and if you are a little lost by now don't worry: this would be expected from a novice physical chemist. I suggest you go back and retrace the solution one more time. Once you fully understand it you should feel good about it. The strategy we used in this problem can be applied to a large number of questions and riddles that involve heat transfer. But wait – you are not done yet; there is a dramatic sequel to the story!

Problem 3.2 | A mountain saga – sequel.

A strong, sudden wind arises and dries the mountain climber's clothes in less than 20 minutes. (B) How much heat has she lost in this process? (C) How much total heat has she lost – rain and wind exposure – in this episode? (D) What should be her body temperature if the metabolism cannot cope with this heat loss?

» Solution – Strategy

Now we have another heat loss, but the "players" are the same: a human body, $m_1 = 65.0$ kg, and rain, $m_2 = 1.2$ kg. They are now at the same temperature, which is $T_1 - \Delta T_1 = 309.75 - 0.55 = 309.20$ K, or $309.20 - 273.15 = 36.05°$C. (A little chilly.)

But this time the heat transfer is of a different kind: the rainwater does not get warmer – it *evaporates*. The rainwater changes phase from liquid to vapor and we call this process a *phase transition*. A phase transition does not come free – water needs a lot of energy to break clusters of molecules in liquid phase and turn them into free, unconnected (sort of) water molecules in vapor phase.

As the phase transition involved in this case requires that liquid water evaporates the heat needed for this phase transition is called *heat of vaporization*, ΔH_{vap}. The values for ΔH_{vap} of many different substances have been determined experimentally and are in textbooks and tables with physical chemical data that could be found online. Heats of vaporization – and other phase transitions – are given per unit of mass or, more commonly, per 1 mol. The heat needed to vaporize 1.2 kg of rainwater soaking the mountain climber's clothes, q_3, depends on ΔH_{vap} per mole and the number of moles of water ($n = 1.2$ kg/0.01802 kg mol$^{-1} = 66.6$ mol) is given as:

$$q_3 = n \times \Delta H_{vap} \tag{3-5}$$

Note that there is no ΔT in this expression for heat transfer as the temperature of water does not change throughout the process. Most sources give ΔH_{vap} of water as 40.65 kJ mol^{-1} so we insert this value into (3-5) and get

$$q_3 = 66.6\,\text{mol} \times 40.65 \times 10^3\,\text{J mol}^{-1} = 2{,}706{,}992\,\text{J} = 2.71\,\text{MJ}$$

This answers question (B): two point seven one megajoules. To answer question (C) you should add q_1 and q_3:

$$q(\text{tot}) = 150{,}960 + 2{,}706{,}992 = 2{,}857{,}952\,\text{J} = 2.86\,\text{MJ}$$

Done for part (C). Now be aware that this is a *large* energy loss. To answer question (D) you need to use (3-1)

$$q = m_1 \times C_p \times \Delta T$$

and calculate ΔT using the expression $\Delta T = q(\text{tot})/(C_p \times m_1)$:

$$\Delta T = 2.86 \times 10^6[\text{J}]/65.0[\text{kg}] \times 4.187 \times 10^3[\text{J K}^{-1}\,\text{kg}^{-1}] = 10.5\,\text{K}$$

Answer (D): If the mountain climber's body did not immediately replenish the lost heat her body temperature would drop by more than ten degrees, $t(\text{body}) = 36.6 - 10.5 = 25.1°$C. If her metabolism did not switch into high gear – long before her temperature drops several degrees Celsius – she would die. To keep the temperature of her body about constant she would have to burn, fast, a lot of food and reserve nutrients in the body. If this has happened as described the mountain climber would suffer from a severe hypothermia and possibly experience disruption of some bodily functions. You may now appreciate the fact we all take for granted today: our clothes are coated with water-repellant materials.

Addendum: What can we do to improve this calculation? I would check the data for human body heat capacity. And, true, the cited data are different than the heat capacity of pure water. When reviewing these data I immediately ran into the following problem: not all human bodies have the same heat capacity. There are heat capacity values for younger people, older people, leaner people, less lean people, healthy people, and not so healthy people. So assuming that a younger, more muscular (i.e., more protein, less fat) female body would fit the description of our mountain climber I chose 3.5 kJ/kg °C [1] as the value for the heat capacity of her body. Perhaps you would select a slightly different value after reading other references [2, 3] but it would still be significantly different from water. So here is a question: Assume the human body heat capacity of 3.5 kJ/kg °C. Estimate the mountain climber's body temperature, after experiencing the heat loss (i.e., soaked in cold rain which then evaporates in strong wind).

Note: You may use (3-2) but you cannot simplify it any further. Why? Because $C_p(1 = \text{body})$ is now 3.5 kJ/kg K and $C_p(2 = \text{water})$ is 4.187 kJ/kg K or, to make the matter simple, just 4.2 kJ/kg K. (Also note that 3.5 kJ/kg °C is the same as 3.5 kJ/kg K, that is, "something/°C" is the same as "something/K." The *change* in temperature of 1°C is the same as the change in temperature of 1 K.)

I suggest the following alternative. Forget the formulas and the calculator and answer the following question: What is the amount of heat transferred to the rainwater, same or different? If you are not sure how to answer it ask yourself this question: What is the total amount of heat in the climber's body before the rain? Keep in mind that the human body heat capacity, 3.5 kJ/kg K, is smaller than 4.2 kJ/kg K, the water heat capacity; this means that a human body contains *less* heat for the same temperature and mass. Put these things together – this is a good little mental workout – and boldly state your estimate.

(A) Lower than 25.1°C – explain:

_____

_____

_____

(B) Higher than 25.1°C – explain:

_____

_____

_____

Here is a little experiment we all have done many times: the beverage in your glass is warm and you add a couple of ice cubes to it and swirl the glass to make the ice cubes melt. How cold is the beverage when the ice melts?

3

| Problem 3.3 | Some like it cool. |

You open a 12 ounce can of your favorite beverage and pour it into a plastic "glass." To cool it down quickly you add two large ice cubes and swirl the glass vigorously. When the ice cubes are completely melted you taste the beverage and like it. What is the temperature, in °C, of the beverage in your glass? The beverage, when poured out of can, was at room temperature and the temperature of ice cubes taken from a freezer was –8°C. Assume that each ice cube is a perfect cube with a side 2.54 cm long.

Answer

T(final) = 16.51°C   (289.66 K)

$T_f(1)$(beverage) = 297.82 K   (For the process: ice at –8°C → ice at 0°C)

$T_f(2)$(beverage) = 291.06 K   (For the process: ice at $T = 273.15$ K (melting) → water at $T = 273.15$ K)

$T_f(3)$(beverage) = 289.66 K   (For the process: icy water at 273.15 K→ water + beverage at 289.66 K)

» Solution – Suggestions and Tips

It is clear what is happening here – a heat transfer. The beverage is at the initial temperature T_1 and the ice is at the initial temperature T_2. The beverage passes heat to the ice cubes; they cool the beverage and melt into water. The cold water mixes with beverage which passes more heat – and cools down some more – while the cold water warms up to a final temperature. This is the same temperature the beverage reaches while cooling down and thermal equilibrium is established.

You may also use the following assumptions:

- The beverage has the same heat capacity as pure water (i.e., ignore the presence of colors, sweeteners, and CO_2 gas in the beverage).
- The heat passed from the beverage to the ice, q_1, will be the same as the heat absorbed by the ice cubes, q_2.
- Ignore the heat transfer from your hand to the plastic glass and the beverage in the glass; also ignore any heat exchange between the beverage and ice with the air in the room.
- Also, make the usual assumption that 1 mL of water (or beverage) equals 1 g.

These are fairly sound assumptions and you are probably already familiar with some of them. If you think further approximations and assumptions are needed, by all means state them. (Learn to use the tricks and devices physical chemists use when solving a problem.) There is one thing that makes this problem less than trivial. The two bodies – the beverage and the "coolant" – exchange heat not in one but in three different processes, or three stages. (Or so we physical chemists divide the problem in order to solve it.) Let us start with a little bookkeeping list:

- First, when you take ice cubes from the freezer their temperature is –8°C. You put them in the beverage at room temperature and the heat from the beverage flows toward ice and warms it up. Note that the ice is still the solid ice; only its temperature was raised from –8°C (265.15 K) to the point where ice starts melting and turning into water, $T = 273.15$ K. But the heat capacity of ice is different from that of liquid water and you will have to insert the correct values. Also, you will have to work a little bit to figure out how much water – I strongly suggest you use moles instead of grams – there is in two ice cubes. Recall what you know from experience: ice floats on water so it must have higher buoyancy – or lower density – than liquid water. Find out what it is. When you finish this part you will find that the beverage has cooled down a little; you may call this temperature $T_f(1)$; and no, it is *not* equal to the ice melting temperature, 0°C. Calculate $T_f(1)$ and write it down.
- Second, now comes the phase transition defined by the following process: the heat from the beverage – now at the temperature $T_f(1)$ – melts the ice which has warmed up to 0°C. During this process the beverage cools down considerably while the ice stays at 0°C – as it should be during a phase transition process. This is the same kind of process as in the "Sequel" to the mountain saga (the rain evaporates while maintaining constant temperature, T_f, in that problem). The beverage has now reached $T_f(2)$, and again no, it is not equal to the ice melting temperature, 0°C. Write down $T_f(2)$.
- Third, from now on it is all downhill; the problem you have to solve is the same as with the mountain climber and cold rain. You have a beverage of mass m_1 (or the number of moles n_1 – if you are working with moles) at the temperature $T_f(2)$ and water from the melted ice cubes (n_2 moles) at temperature 0°C. The heat flows from the beverage, cools it even more, and warms up the cold water. The temperature of the beverage will fall down from $T_f(2)$ to the equilibrium temperature $T_f(3)$. The temperature of the cold water from the melted ice cubes will warm up from 0°C to $T_f(3)$ and thermal equilibrium will be established. Find $T_f(3)$ the same way we have found T_f in the problem with the rain-soaked mountain climber.

So the answer is $T_f(3)$, in °C. If you have got the right answer (check above) you should congratulate yourself – you have pretty much mastered the tricks of energy balancing between two bodies at different temperatures (which happens all the time to all the bodies around us).

A comment: Next time you have a glass of beverage with your friends think of how much more you know and understand what is happening, right there, right in their glasses. What –are you going to take a cold beverage from a fridge? Then you may want to think about what happens to your vocal cords, $T = 36.6°C$, when you swallow 355 mL of cold, $T \sim 4°C$, liquid.

A note on open, closed, and isolated systems: Let us think for a moment of these two examples. In Problem 3.1 the mountain climber passes heat, that is, energy, to cold, rain-soaked clothes. The mountain climber is a system; so are the rain-soaked clothes. They exchange energy only and we call such systems *closed* systems. In the second example the ice and the beverage exchange energy too. But they also exchange much more than that – the ice cubes melt and mix with the beverage. We say they also exchange mass. A system that exchanges both energy and mass is called an *open* system. You and I and every living organism are open systems. Think of the motions of your body (energy exchange) and the inhaling and exhaling (mass exchange), and you will see why this is so. There is also a third type of system. In technical physics and thermodynamics there is an approximation known as an *isolated* system: a system that does not exchange anything with its environment. In reality there are no truly isolated systems.

Make a note: *Open, closed, and isolated systems.*

References

Human Body Temperature
1. Giering K, Lamprecht I, Minet O, Handke A (1995) Determination of the specific heat capacity of healthy and tumorous human tissue. Thermochim Acta, 251:199–205
2. Jay O, Gariépy LM, Reardon FD, Webb P, Ducharme MB, Ramsey T, Kenny GP (2007) A three-compartment thermometry model for the improved estimation of changes in body heat content. Am J Physiol Regul Comp Physiol 292:R167–R175
3. URL: http://www.engineeringtoolbox.com, Accessed July 2009

4   Thermodynamics

I mentioned above that thermodynamics is concerned with changes in the total energy, E, of the system. For example, you help your friend carry a piano to the third floor. While doing this you give away mechanical work (a lot of it, in this case); we will label it as $-w$. You also get really hot while carrying the piano and exchange your body heat with the environment (you sweat); we label it by $-q$. The total energy change in your body is a sum of all the work and all the heat exchanged and we express this by the following equation: $E = -w - q$. As you are giving away both the work and the heat their signs are negative. If, however, someone has carried you up to the third floor and then you drank a cup of hot tea, the sign for the work done and the heat exchanged will be plus – from your point of view – and the energy equation will read $E = +w + q$. In thermodynamics, we call the total energy of a system as *internal energy*, U, or more often *enthalpy*, H:

At constant volume: $U = w + q$ (4-1)

At constant pressure: $H = w + q$ (4-2)

A note on the end and middle points: Much earlier I said that work, w, depends on path; so does heat exchange. We say that work and heat are *path functions*. (You may think of your life – it depends on the path you take.) This is not the case with the *total energy* of a system. In spite of its name there is nothing dynamic in thermodynamics; it is more an accounting of the events that have already happened. The thermodynamic functions, or state functions – U, H, S, G, A – account for the amount of work, heat, and other forms of energy exchanged. They do not account for the order or the timing of these activities. When we say energy or enthalpy we actually mean a *change* in energy and a change in enthalpy. We cannot measure energy or enthalpy (or any other thermodynamic function) at any single point along the path but we can measure how much they have changed between two points along the path; usually these are the beginning and the end. This is why we call the thermodynamic functions *point functions*; they depend on the final and the initial points. So I should write the expression given above as $\Delta H = -q - w$. Unfortunately physical chemists

P.-P. Ilich, *Selected Problems in Physical Chemistry*,
DOI 10.1007/978-3-642-04327-7_4, © Springer-Verlag Berlin Heidelberg 2010

are often sloppy and keep writing U, H, S when they should be writing ΔU, ΔH, ΔS, and this is not helping people like you.

Let us make two little notes: (1) Work and heat exchange depend on *path*. (2) Energy, U, enthalpy, H, entropy, S, and Gibbs energy, G, depend on the difference between two *points* only: usually the final state and the initial state.

A note on thermochemistry and chemical bond enthalpies: Everything that happens in chemistry can be accounted for through energy balancing. You can do this balancing if you know the energies of chemical bonds. This is like with words and sentences – first you have to know the letters. How do you obtain a measure of the energy of a chemical bond? In the past, it was done like this: first, you conduct a chemical reaction and measure the total energy change. For example, you take methane, CH_4, add dioxygen, O_2, and burn methane to form water, H_2O, and carbon dioxide, CO_2. You carefully measure the heat, the enthalpy change during this reaction. Then you divide this enthalpy by the number of bonds broken (C–H) and bonds made (C–O), (H–O), and the number you get is an approximate *bond enthalpy* for this reaction. If you think a little about this you can see that it takes a lot of different experiments to find the exact enthalpy of, for example, the C–H bond only. For this reason, thermochemistry – the branch of physical chemistry that was concerned with such experiments – was probably the most important part of the whole physical chemistry. This has radically changed over the past 25 years. Today, you will use *molecular modeling* software to calculate enthalpy, internal energy, and other thermodynamic properties of atoms, molecules, and ions. The resulting bond enthalpies, obtained in few minutes or few seconds of calculations, are equal to or more accurate than the bond enthalpies obtained by thermochemical experiments carried out in the course of several hours, days, or sometimes weeks. Today, the topic of thermochemistry and chemical bond enthalpies is better taught as laboratory exercises in *computational physical chemistry*.

4.1 Entropy

Summing up all the changes that accompany heat transfer between two bodies leads us to the first law of thermodynamics. The law posits that energy never gets destroyed – it just passes from one body to another; the law also posits that energy cannot be created anew. This type of observation has been made before the invention of thermodynamics: The Greek philosopher Anaxagoras stated some 2,500 years ago that "You get nothing from nothing." So, the next time you are promised a free lunch, think of the first law of thermodynamics, or think of what smart Greeks knew already 2,500 years ago.

In processes where there is no mechanical or gas expansion work – known as the $p\Delta V$ work – the change in enthalpy equals the amount of heat transferred. This is the case with most living systems. It is customary to use both words – heat and enthalpy – to say the same thing.

The empirical observation that heat always flows from a warmer toward a colder body provides the basis for the second law of thermodynamics. Another thermodynamic function – *entropy* – has been introduced to provide a more complete description of physical and chemical changes of a system. What is entropy? It is defined as the amount of heat divided by temperature: $S = q/T$. A look at its unit – joule per kelvin – tells us that entropy is related to energy, scaled by temperature. The *change* in entropy tells us something more. Let me recall the rain-soaked mountain climber to show you what I have in mind:

(1) In the first event, the climber, with the body temperature $T_1 = 309.8$ K, is in contact with cold rain and loses heat, $q = -151$ kJ. So the entropy change of the climber will be given by the ratio of this heat (note the sign) and the body temperature:

$$\Delta S(\text{climber}) = q/T_1 = -151 \times 10^3 \text{ J}/309.2 \text{ K} = -4.874 \times 10^2 \text{ J K}^{-1}$$

(2) The 1.2 kg of the cold water at $T_2 = 279.2$ K receives this heat, $q = +151$ kJ, which will cause its entropy to change too:

$$\Delta S(\text{rainwater}) = q/T_2 = +151 \times 10^3 \text{ J}/279.2 \text{ K} = +5.408 \times 10^2 \text{ J K}^{-1}$$

So the total change in entropy is the sum of all entropy changes; the entropy change of the climber in contact with cold rain-soaked clothes and the entropy change of the cold rain in the wet clothes:

$$\Delta S_{\text{tot}} = \Delta S(\text{climber}) + \Delta S(\text{rainwater}) = -4.87 \times 10^2 \text{J K}^{-1} + 5.41 \times 10^2 \text{J K}^{-1}$$
$$= +53.42 \text{ J K}^{-1}$$

The total entropy change is *positive*. Make a note of this.

A note on probabilities: There is a reason for this and the reason is in the way the atoms and molecules in our bodies and the objects around us are arranged. Think of a simple example, like an everyday event in a household.

>> **Example 4.1**
A cookie in a jar (Figs. 4.1 and 4.2).

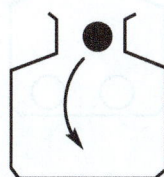

☐ **Fig. 4.1** You place a cookie in a jar

4

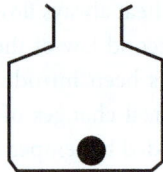

■ **Fig. 4.2** There is *one* cookie in a jar

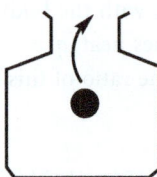

■ **Fig. 4.3** You take the cookie out of the jar

When you reach in the jar the chance (probability P) that you will find a cookie there is 100%. You state this by writing $P = 1.0$ (Fig. 4.3).

» **Example 4.2**
Two cookies in a jar (Figs. 4.4–4.6)

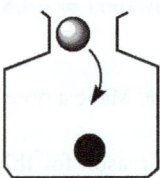

■ **Fig. 4.4** You put one peanut butter cookie and one chocolate chip cookie in a jar

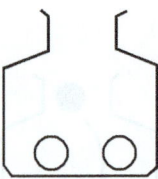

■ **Fig. 4.5** There are two different cookies in the jar – but you cannot see them

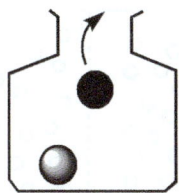

Fig. 4.6 You reach into the jar to take out a cookie

The cookie you take out could be the peanut butter cookie or it may not be (Fig. 4.6); the chance is 50% and you write $P = 0.5$. Let us retrace what you have done. First, when you put one peanut butter cookie in the jar you knew, *for certain*, $P = 1.0$, this is the cookie you wanted. But when you reach into the jar with two different cookies (Figs. 4.5 and 4.6) you are not certain anymore. The cookie you grabbed could be the peanut butter cookie – which you wanted – or it could be another kind of cookie. Something new has happened between Example 4.1 and Example 4.2 – an *uncertainty* has entered the system. The more objects you have the more uncertainty there is. Entropy has been defined by Ludwig Boltzmann, an Austrian physicist, as a direct measure of such uncertainty:

$$S = -k_B \ln P \tag{4-3}$$

Here, P is the total probability of doing something in the way you want it to happen. The k_B term is the so-called Boltzmann constant, a proportionality factor. There is no entropy for a system of one body, one choice, one possibility. But already for *two* bodies the probability is smaller than 1.0. The logarithm of the probability is smaller than zero and the *entropy is positive*.

One or two cookies may not seem as a best way to explain a thermodynamic function so let us try another example: exchange of heat between a hot and a cold body. Imagine two bodies at different temperatures (Fig. 4.7), a hot body (gray circles) and a cold body (open circles). Think of the hot body as made of small *hot particles* and the cold body as made of *cold particles*. There are a lot of these particles, likely in the range of Avogadro's number, $N_A = 6 \times 10^{23} \text{ mol}^{-1}$. Now the two bodies come into contact (Fig. 4.8).

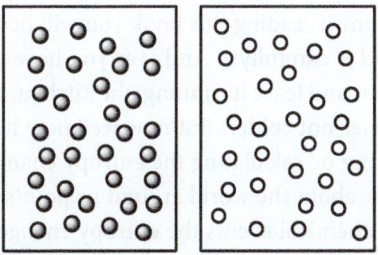

Fig. 4.7 Hot (gray) and cold (open) bodies, separated

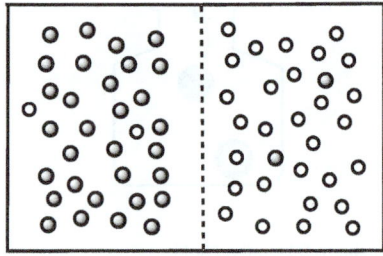

🔲 **Fig. 4.8** Hot and cold bodies in contact. Particles start passing from one side to another

So the same old story: hot becomes colder and not the other way around. And why exactly is this happening? Because once the hot and cold particles get mixed (Fig. 4.9) the chance of finding each of them and putting them back into separate chambers is proportional to $(2 \times 10^{-23})^{6 \times 10^{\cdot}23}$, or a number with 10^{529} zeros in front of it. The temperature of the system is another important factor and we state it in this way: *Since the hot particles are moving faster the chance of finding a hot particle is even lower.* Think of this number for a moment, the probability of heat flowing from a cold body toward a hot body: $1/10^{529}$. For comparison, one second of time is about $1/10^{9}$ part of your lifetime. You will need another 10^{58} lifetimes to make one, perhaps correct, selection.

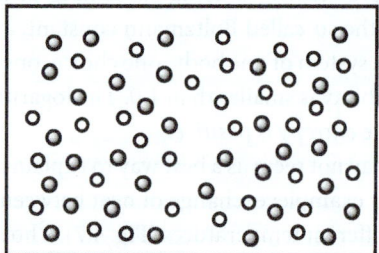

🔲 **Fig. 4.9** In the end all particles are mixed and at the same temperature

So, yes, it is possible to do this, but in practice the probability for doing this is too low. For the same reason Monday comes after Sunday and not the other way around, and when you finish reading this book you will be a little older than when you started it. (So – read it carefully!). And – as you have certainly heard – if you open the refrigerator door and leave it running, the kitchen, when you return several hours later, will be *warmer* not colder. But now we know how to account for these occurrences: by measuring or calculating the entropy change of a system. It seems that everything we know about the world around us points to an entropy increase. For many physical and chemical events the entropy change is usually small but in some natural processes it can reach truly large scales. Such is the entropy increase in case of mixing of large masses of gases.

Problem 4.1 | The not so good old times.

The atmosphere – about 5.1480×10^{18} kg of a gas mixture [1] composed of 79% N_2 (v/v) and 21% O_2, if we ignore Ar, CO_2, water vapor, and, at certain places particularly, pollutants – has not always been like this. Today we know that about 3.5×10^9 Gyr (Gyr = gigayear = 10^9 year) ago the Earth atmosphere was mostly dinitrogen, N_2. (The initial, small amounts of O_2 – released through volcanic activity and chemical processes – spawned the occurrence of aerobic living organisms, capable of carrying out photosynthesis and converting H_2O to O_2. Most of the dioxygen in the atmosphere today has been created through plant photosynthetic activity; check Problem 12.9.) Assuming that N_2 was constant throughout this time calculate the total entropy of mixing of N_2 and O_2 accompanying the formation of the present atmosphere.

» Solution – Strategy and Calculation

The strategy in this problem is straightforward: we assume the ideal behavior of N_2 and O_2 and calculate the entropy of mixing using the standard formula:

$$\Delta S = -nR(x_1 \ln x_1 + x_2 \ln x_2) \tag{4-4}$$

The molar fractions are known – they are the same as the volume fractions in the mixture of two ideal gases: 0.79 for N_2 and 0.21 for O_2. What we do not know, however, is n, the total number of moles of the two gases in the Earth atmosphere today. Given that molecular masses of the two gases are different their mass ratio (w/w) is different from the volume ratio (v/v). We will simplify this problem by using *little big trick* #4: use 1 mol as a default value when no quantity is given. Ask yourself a question: What is the mass of *one mole* of air? And the answer is, It is the mass of 0.79 moles of N_2 with a mass of 28.014 g mol^{-1} and 0.21 moles of O_2 with a mass of 31.999 g mol^{-1}. Translate the last sentence to a formula and calculate the molar mass of air (a somewhat simplified air):

$$m(\text{air}) = 0.79 \times 28.014 + 0.21 \times 31.999 = 28.85 \text{ g mol}^{-1} = 2.885 \times 10^{-2} \text{kg mol}^{-1}$$

Note that this is a good number for calculations in this kind of problems although conceptually it means nothing: there is no molecule of air. From this number and the data for the total mass of the Earth atmosphere we will get the total number of moles of the gases (do not forget to convert the molar mass of air to kilograms):

$$n(\text{atmosphere}) = 5.1480 \times 10^{18} \text{ kg}/2.885 \times 10^{-2} \text{kg mol}^{-1} = 1.784 \times 10^{20} \text{mol}$$

Punch few more keys on your calculator to get ΔS, the entropy of mixing of N_2 and O_2 in the Earth atmosphere:

$$\Delta S = -1.784 \times 10^{20} \text{mol} \times 8.314 \, \text{J K}^{-1} \text{mol}^{-1} (0.79 \ln 0.79 + 0.21 \ln 0.21)$$

$$\Delta S = -1.483 \times 10^{18} [\text{J K}^{-1} \, \text{mol}^{-1}] \times -0.514 = +7.62 \times 10^{20} \, \text{J K}^{-1} \, \text{mol}^{-1}$$

Comment: This is a large, positive entropy, probably about as large a number for entropy as you will ever see (unless, that is, we consider tackling the entropy change related to the expansion of the Universe). Also note that entropy of mixing, no matter how you define the problem, always has to be positive. The logarithms of the molar fractions, always smaller than one, are negative and the overall sign is negative. So there! In part, the equation is designed in order to express the problem we mentioned before: how to separate two cookies. Only now there are $(7.6 \times 10^{20}) \times (6 \times 10^{23}) \approx 5 \times 10^{43}$ "cookies" – a lot to choose from.

4.2 Gibbs Free Energy

We have seen above how entropy is created when heat flows from a hotter to a colder body. Once created, this entropy cannot be destroyed or reversed; it is a function of the energy lost in the process. (You may not notice this but, once you have finished helping your friend move to a new apartment, your body is not the same anymore; it will never be.) The first law of thermodynamics gives us a measure of the total energy of a system. The second law of thermodynamics tells us that a certain amount of energy is lost during the transfer. You need to know how much of the total energy is actually available. There is a thermodynamic function that tells us this: it is called *Gibbs free energy*, G, in honor of Josiah Willard Gibbs, an American physicist. The G function, defined through H and S, reads

$$G = H - TS \tag{4-5}$$

The change in G, that is, the difference in G between G now and G yesterday, is given as

$$\Delta G = \Delta H - T\Delta S \tag{4-6}$$

This is probably the *most important thermodynamic equation*. The change in Gibbs energy, ΔG, depends not only on chemical change – like in a chemical reaction – but also on physical parameters, like temperature and pressure. J. W. Gibbs was particularly involved with studying the cases where ΔG is caused by physical changes. Along this line, the following little riddle involves an ice skating rink, ice, water, and a skater.

Problem 4.2 | Remember Tonya Harding?

Figure skating blades are 18–28 cm long but due to their slight curvature ("rocker") only about 6.0 cm of a blade is in contact with ice. Similarly, due to its inner curvature ("the bite angle") only about 1.0 mm of the total blade width (~4 mm) is touching the ice. Most of the time a skater, mass = 61.5 kg, is standing – and exerting a pressure on the surface – on one skate only. The surface of the skating rink at this temperature and pressure, $t = -1°C$, $p = 101{,}325$ Pa, is ice. Given that the densities of water and ice under these conditions are 0.999 g cm^{-3} and 0.917 g cm^{-3}, respectively, calculate how much will Gibbs free energy change under the skater's blade for water and how much for ice. Use these ΔG values to decide whether it is easier to skate on a solid ice or on (a thin layer of) water.

» Solution – Strategy and Calculation

When a problem looks difficult you should try to break it into smaller parts and take a look at each of them. This may be easier said than done as you probably do not know how to break a problem into smaller pieces. This is an art – and we will master it together. Making a list, like a list of the words mentioned, or a list of the quantities and numbers given, or a list of the questions asked, can help us move ahead.

What words are mentioned above?

- An ice skater, a single skate blade, ice, water, pressure, temperature, and ΔG.

What numbers are given?

- 61.5 kg for the skater's mass, 6.0 cm [6.0×10^{-2} m] by 1.0 mm [1.0×10^{-3} m] for the skate's contact area, and $-1°C$ for the ice–water temperature.

And what is the question?

- How much will the chemical potentials of ice and of water change from when it is exposed to normal air pressure to when it is exposed to the additional pressure by the skater?

This question may look a little convoluted and we will translate it to symbols and formulas. We have *chemical potential*, that is, G per one mole, G_m (or μ), and we also have a change; this means ΔG_m. Since both ice and water are present on the skating surface we will have to account for changes of two chemical potentials: ΔG_m(ice) and ΔG_m(water). Not bad for a start.

The riddle states that the changes in G are due to the change in pressure. Check your textbook and under the section "Dependence on Gibbs Energy on Temperature and Pressure" you will find this formula:

$$\Delta G = V \times \Delta p \qquad (4\text{-}7)$$

The difference, Δ, in (4-7) means the following: G(final state) minus G(initial state). The same applies to the pressure change. The previous equation can then be written as

$$G_f - G_i = V \times (p_f - p_i)$$

And what are the *initial* and *final* states here? Do not ponder too much at this point. The initial and final need not mean a truly initial (that is, the beginning of something) and the final states (the end of everything); we can simply designate them as state #1 and state #2. The text of the riddle says

(1) the ice at normal pressure and
(2) the same ice but with the skater standing on it

Clearly, the initial pressure, p_i, is the pressure of air, p(air), and the final pressure is the pressure of the air and the skater standing on the ice, $p_f = p$(air) + p(skater). The *difference* in pressure is given by the following expression:

$$\Delta p = [p(air) + p(skater)] - p(air) = p(skater)$$

Let us calculate the pressure of the skater standing on one skate. Pressure, p, is force divided by area, F/A. The force is the skater's weight: notice *weight*, not mass, that is, the mass multiplied by gravity:

$$F = 61.5[kg] \times 9.81 \, [m\,s^{-2}]$$

$$F = 6.03 \times 10^2 \, kg\,m\,s^{-2} = 6.03 \times 10^2 \, N$$

You can calculate the contact area of the skate by multiplying the average contact length of the blade with the average contact width of the blade edge:

$$A = 6.0 \times 10^{-2} \, [m] \times 1.0 \times 10^{-3} \, [m] = 6.0 \times 10^{-5} \, m^2$$

So the pressure the skater exerts on the ice is given as (2-1):

$$\Delta p = F/A$$

$$\Delta p = 6.03 \times 10^2 \, [kg\,m\,s^{-2}]/6.0 \times 10^{-5} [m^2] = 1.0 \times 10^7 \, kg\,m^{-1}\,s^{-2}$$

$$= 1.0 \times 10^7 \, Pa \, (pascal = newton/m)^2$$

Now go back to (4-7). Since the question mentions a "change per mole" you should re-write the equation by adding "m" as subscripts to the Gibbs energy and volume:

$$\Delta G_m = V_m \times \Delta p \tag{4-8}$$

What is V_m here? It is the volume of one mole; *one mole of ice*. You will find the V_m from the ratio of the mass of one mole of ice and its density; the latter is given as $0.917 \, g \, cm^{-3}$. Insert these values and you will get

$$V_m(\text{ice}) = 18.02 \text{ g mol}^{-1}/0.917 \text{ g cm}^{-3} = 19.65 \text{ cm}^3 = 1.97 \times 10^{-5} \text{ m}^3$$

You can now calculate the Gibbs energy change from the molar volume and the pressure:

$$\Delta G_m(\text{ice}) = 1.97 \times 10^{-5} \text{ [m}^3\text{]} \times 1.0 \times 10^7 \text{ [Pa]}$$

Let us re-check the units: $1 \text{ Pa} = \text{N/m}^2 = \text{kg m s}^{-2}/\text{m}^2 = \text{kg m}^{-1} \text{ s}^{-2}$ and $\text{Pa} \times \text{m}^3 = \text{kg m}^2 \text{ s}^{-2} = \text{joule, J}$, so the change in the Gibbs energy of ice when the skater is standing on it, rounded up to the closest value, will be

$$\Delta G_m(\text{ice}) = 197 \text{[m}^3\text{]} \times \text{[kg m}^{-1} \text{ s}^{-2}\text{]} \approx 200 \text{ J}$$

You will get the molar volume for water the same way – as a ratio of the molar mass and density:

$$V_m(\text{water}) = 18.02 \text{ g mol}^{-1}/0.999 \text{ g cm}^{-3} = 18.04 \text{ cm}^3 = 1.804 \times 10^{-5} \text{ m}^3$$

And the change in the Gibbs energy of one mole (the change of chemical potential) is given as

$$\Delta G_m(\text{water}) = 1.804 \times 10^{-5} \text{ [m}^3\text{]} \times 1.0 \times 10^7 \text{ [kg m}^{-1}\text{s}^{-2}\text{]} = 1.8 \times 10^2 \text{ J}$$

Fine – but what are these two numbers, 200 J and 180 J, telling us? Read the following note.

A note on higher and lower energies: Once upon a time physicists got together and decided the following: a *smaller energy* means a *more stable* system. But beware: by smaller is meant *smaller positive* number; it also means a *larger negative* number. The same applies to changes, processes, and events. The change in total energy which is negative, that is, it goes from a larger positive to a smaller positive number, means that the process will occur spontaneously. The system "likes" this change. (This statement definitely does not extend to monetary matters: A loss of money cannot be good.)

We have obtained, after breaking the problem into smaller parts and carefully progressing from one step to another, two energies: 200 J and 180 J. The 180 J is a smaller number, so this process, this event, is more likely to happen. And which event is that? Skating, that is, pressing with your skate on the (thin layer of) *liquid water*. So when the skater steps on ice, the ice will melt to reduce the change in Gibbs free energy from 200 to 180 J. The skater is skating on water. She moves away and the water freezes back to ice – for the temperature in the skating rink is –1°C. Riddle solved. This has been suspected for some time; only now we have shown it by a number. Let us not forget why we are reading this book: physical chemistry will help us understand the nature of the things around us and the way they change.

A comment: Temperature is another parameter that causes changes in all thermodynamic functions, entropy and Gibbs energy included. This is particularly true for systems which do not exchange the so-called $p\Delta V$ work, the work resulting from expansion or compression of a gas. When calculating the changes in entropy and Gibbs free energy, caused by a slow change in temperature, we assume that these processes are reversible. We do this for the sake of consistency with the classical, technical thermodynamic relations, based on crude heat and the gas expansion/compression work measurements. At the same time we know that at the molecular level a complete reversibility is practically impossible.

Problem 4.3 | **Josiah Willard Gibbs in emergency room.**

A male patient, mass $= 75.3$ kg, develops acute postoperative fever. Within an hour, his body temperature rises from the normal level, $t_1 = 36.6°C$, to $t_2 = 41.2°C$ before it gets stabilized by a massive dose of antipyretics. Out of curiosity you wonder how much his energy consumption has increased during this period. Calculate the changes in entropy and Gibbs free energy of the patient assuming that his body can be approximated with 75.3 kg of pure water.

» Solution – Strategy and Calculation

The question here is, How much do the energy and entropy of a certain amount of water change when I warm it up from T_1 to T_2? No chemistry, no solution and mixing, and no gas expansion are present in this process. I suggest you start with the formula connecting G, H, and S:

$$G = H - TS \tag{4-9}$$

So, when the "process," that is, increase in the body temperature, is finished the *change* of G will equal the following difference:

$$\Delta G = G_2 - G_1$$

$$\Delta G = \Delta H - \Delta(TS) \tag{4-10}$$

The change in enthalpy, ΔH, as you may find in a physical chemistry textbook, under the section on first law of thermodynamics, equals q, the heat. Check Problem 3.1, with the mountain climber in cold rain, and you will find that heat is expressed as a product of heat capacity, the amount of the substance, and the difference in temperatures:

$$\Delta H = q = C_p \times m \times (T_2 - T_1)$$

Note that the amount in the previous equation is given as a mass, m; it can as well be given as the number of moles, n.

What about the $\Delta(TS)$ term? This is a little tricky. You have to understand that Δ is a *differential* operator. There is a rule in calculus about it and it goes like this:

$$\Delta(TS) = \Delta T \times S + T \times \Delta S \tag{4-11}$$

We will use this expression to calculate the total entropy change of the unfortunate Josiah.

» Solution – Calculation

(A) Enthalpy change:

$$\Delta H = C_p \times m \times (T_2 - T_1)$$
$$= 4.187 \times 10^3 \, \text{J} \, \text{K}^{-1} \, \text{kg}^{-1} \times 75.3 \, \text{kg} \times (314.4 - 309.8) \, \text{K}$$
$$= 1.45 \times 10^6 \, \text{J}$$
$$= 1.45 \, \text{MJ}$$

(B) Entropy change:

We have two terms, $\Delta T \times S$ and $T \times \Delta S$. We have to think a little bit about these two terms and use common sense (and a little bit of knowledge). The ΔT term is easy – the difference between two temperatures, $T_2 = 314.4$ K and $T_1 = 309.8$ K. But what is S? It is the entropy content at temperature T, that is, a single-point value for the entropy function. Previously, on several occasions I said that in general you cannot know the entropy content (or the enthalpy content or the Gibbs energy content) at a single point; only a *difference* between two points can be determined. And true to this statement most current physical chemistry textbooks will tell you that S is not defined, therefore G is not defined and you cannot calculate it this way. But physical chemists should know better: read the following short note.

A note on the third law of thermodynamics: There is a point of lowest possible temperature, close to zero degrees kelvin, $T_0 \approx 0$ K, and for every solid body that forms a perfectly regular crystal the entropy at T_0 equals zero. Water can crystallize so at 0 K the entropy of water is zero (or nearly so). It is perhaps a little stretch to assume that Josiah Gibbs would make a perfect crystal, even at 0 K, but we will not be too pedantic about this point right now. So what is the entropy of water at, say, $T_1 = 309.8$ K ? It equals the *difference* between the entropy of water at 309.8 K and the entropy of water at 0 K, which is zero. We can calculate this or we can take the listed value: it is determined for room temperature, $T = 298.15$ K, and is listed as the *absolute entropy* or *standard molar entropy*. For water at room temperature, $S_{298} = 69.9 \, \text{J} \, \text{K}^{-1} \, \text{mol}^{-1}$. True, $T = 298.2$ K is not the same as $T_1 = 309.8$ K, but we should not worry about this detail now; the error is likely to be less than 1%. The entropy of water will change significantly between 273 and 274 K, when ice (solid) melts into water (liquid) because there is a large increase in the disorder of the system, and very little between 298.2 and 309.8 K, so we can ignore this error now. (I would like to hope that by this point you are beginning to agree with some of my assumptions, and if you are thinking up of some of your own assumptions by all means write them down!)

Let us plug this little information into our problem. You need to convert Josiah's 75.3 kg of water into moles of water: divide his mass with a mass of 1 mol of water (make sure all masses are given in kg):

$$n = 75.3\,[\text{kg}] \times 1{,}000\,[\text{g kg}^{-1}]/18.02\,[\text{g mol}^{-1}] = 4.18 \times 10^3\,\text{mol}$$

So the first entropy term will read

$$\Delta T \times S = (314.4 - 309.8)[\text{K}] \times 69.9\,[\text{J K}^{-1}\,\text{mol}^{-1}] \times 4.18 \times 10^3\,[\text{mol}]$$
$$= 1.346 \times 10^6\,\text{J} = 1.35\,\text{MJ}$$

The next entropy term, $T \times \Delta S$, you can find in all physical chemistry textbooks. First, we re-write it using small differences – differentials, d:

$$dS = m \times C_{\text{p}} \times dT/T \tag{4-12}$$

When multiplied by T the whole expression takes a much simpler form:

$$TdS = m \times C_{\text{p}} \times dT$$

When this expression is integrated (or summed, if you are taking small but measurable steps) you get

$$T\Delta S = m \times C_{\text{p}} \times \Delta T \tag{4-13}$$

ΔS is an integral (a sum) of all small increments in entropy as the temperature rises from T_1 to T_2. Now insert the known and the given numbers, punch a few keys on your calculator, and you will get

$$T\Delta S = C_{\text{p}} \times m \times (T_2 - T_1) = 4.187 \times 10^3\,\text{J K}^{-1}\,\text{kg}^{-1} \times 75.3\,\text{kg} \times (314.4 - 309.8)\,\text{K}$$
$$= 1.45 \times 10^6\,\text{J} = 1.45\,\text{MJ}$$

Note that you get the same expression – and the number – as for the enthalpy change, ΔH. This is stated in textbooks as a common truth but rarely derived.

The entropy change in a body of 75.3 kg of water, when its temperature is raised from 36.6 to 41.2°C, is calculated using the formula for the "reversible" change in entropy with temperature change. We will again use the differential "d," which is as small change – in both temperature and in entropy – as we desire it to be. We write

$$dS = m \times C_{\text{p}} \times dT/T$$

It follows from calculus that the integral (sum of small parts) of the expression dT/T is a logarithmic function of temperature, $\ln T$. We evaluate it between the two boundaries T_2 and T_1:

$$\Delta S = m \times C_p \times (\ln T_2 - \ln T_1) = m \times C_p \times \ln(T_2/T_1)$$

Insert the numbers for the mass [kg], heat capacity [J kg^{-1}], and the two temperatures [K] and you will get

$$\Delta S = 75.3 \,[\text{kg}] \times 4.187 \times 10^3 \,[\text{J kg}^{-1}] \times \ln(314.4/309.8)$$

$$\Delta S = 4.65 \times 10^3 \,\text{J K}^{-1}$$

This is a big number but, again, you see that it is hundreds of times smaller than the change in enthalpy – because you divide the energy change by the temperature. And this is the answer to the first question in the riddle.

Now for the second question, the ΔG. But let us first check the signs of all the terms. The heat is passed to the body, therefore the enthalpy change is positive. Both the enthalpy and entropy change terms are positive. You may use the expressions (4-9) and (4-11) and insert the numbers for ΔH, $\Delta T \times S$ and $T \times \Delta S$:

$$\Delta G = \Delta H - \Delta(TS) = \Delta H - \Delta T \times S - T \times \Delta S$$

$$\Delta G = 1.45 \,\text{MJ} - 1.35 \,\text{MJ} - 1.45 \,\text{MJ} = -1.35 \,\text{MJ}$$

This is the total change in the Gibbs energy and the answer to the second question in the riddle. We should ask the following question: Where is this large negative change in Gibbs energy coming from? If you look at Josiah's body as a bag of water then he has *gained* energy. But if you go deeper, just a little deeper, you will have to consider the fact that a human body is a lot more than a bag of water; it contains innumerable other small and large molecules and it is running myriads of biochemical reactions every moment of the time. It is these biochemical, metabolic reactions that kicked in and produced a lot more energy than under normal circumstances. In fact they outputted 1.34 MJ of extra energy to raise the body temperature as a response to an infectious process in progress. So, yes, the total body mass has gained energy but the biochemical organism has lost this energy. And, this is a lot of energy. For comparison, this energy expenditure is equivalent to two full weeks – day and night – of light running (6 min/km) for a person of this mass. No wonder a disease, even a mild one, makes you very tired. But, hey, we have saved J.W. Gibbs' life!

Reference

Earth Atmosphere
1. Trenberth KE, Smith L (2004) The mass of the atmosphere: A constraint on global analyses. J Climate, 18:864–875

Table II

Summary of basic thermodynamics – what have we learned?

Write down the new words and their English descriptions. Then re-write these sentences using symbols and numbers.

Words and Phrases	Symbols, Formulas and Numbers
Internal energy, U	$U = q + w$
Enthalpy, H	$H = q + w + p \times V = U + n \times R \times T$
Work due to gas expansion or compression	$w = p \times (V_{final} - V_{initial}) = p \times \Delta V$
Heat and heat content	$q = m \times C \times \Delta T$
Heat capacity, C	C_p – specific heat at constant pressure C_v – specific heat at constant volume
Heat transfer equilibrium	$q_1 = q_2; m_1 \times C_p(1) \times \Delta T_1 = m_2 \times C_p(2) \times \Delta T_2$
Entropy, thermodynamic definition Entropy, probabilistic definition	$\Delta S = q_{reversible}/T$ $S = -k_B \times \ln(\text{Probability})$
Entropy dependence on temperature	$\Delta S = m \times C \times \ln(T_{fin} - T_{init}) = m \times C \times \Delta T$
Entropy of mixing	$\Delta S = -n_{tot} R (x_1 \ln x_1 + x_2 \ln x_2)$ $n_{tot} = n_1 + n_2; x_1 = n_1/n_{tot}$
Gibbs free energy, Change of Gibbs energy	$G = H - T \times S;$ $\Delta G = G_{fin} - G_{init}; \Delta G = \Delta H - T \times \Delta S$
Gibbs energy of mixing	$\Delta G = n_{tot} \times R \times T \times (x_1 \ln x_1 + x_2 \ln x_2)$
Change of Gibbs energy with pressure	$\Delta G = V_m \times \Delta p$
Change of Gibbs energy with temperature	$\Delta G = -S \times \Delta T$

Part III

Mixtures and Chemical Thermodynamics

5   Mixtures and Solutions 53

6   Chemical Reactions and Gibbs
     Free Energy 61

7   Gibbs Free Energy and Chemical
     Equilibria 65

5   Mixtures and Solutions

The world around us is made not of pure substances but of *mixtures*. The air is a mixture of dinitrogen and dioxygen and small amounts of other gases and vapors. The blood flowing through our veins and arteries is a mixture of water and many thousands of other molecules. When we add salt, NaCl, to water and let it dissolve we call this mixture a *solution*. The solid NaCl is the *solute* and the liquid water is the *solvent*. We measure salt in grams (or kilograms) and water in milliliters (or liters). Another way to express the amount of both the solute and the solvent is to convert them to moles. The ratio of the amount of salt to the amount of water is a fixed property for each solution; we call it *concentration*. Like any other ratio, concentration is constant throughout the solution: every single, smallest drop of a solution has the same ratio of salt and water. The same with your coffee, only stir that sugar at the bottom of the cup well so it *all* dissolves. Concentration is therefore an *intensive* property; in general, two concentrations do not add.

| Problem 5.1 | A real emergency. |

On your first day as a volunteer in a refugee camp in one of the troubled areas around the world today, you encounter the following situation: a paramedic assisting an injured local asks you to help him by quickly preparing "1–2 L" of *isotonic saline* solution. "Sure, in a moment" you say, while a little panicked as you walk toward your tent. But there in the tent you discover 1 L bottles sealed and filled with clean water and after few seconds of mentally reviewing your inventory you recall a larger bottle with sodium chloride tablets among your basic medical supplies. (The ones sold in the USA come with a label "Sodium Chloride Tablets 1 Gm, USP Normal Salt Tablets".)

>> **Solution**

OK – you say to yourself, while composing your thoughts – isotonic saline solution, that is, 0.9% solution of sodium chloride, NaCl, in water. And – from the days of

P.-P. Ilich, *Selected Problems in Physical Chemistry*,
DOI 10.1007/978-3-642-04327-7_5, © Springer-Verlag Berlin Heidelberg 2010

your high school/introductory college chemistry – you recall that a foolproof way to prepare a 0.9% solution is to weigh 0.9 g of "something" and dissolve it in so much water to make a 100 mL solution. However, there is no way you can weigh 0.9 g or measure 100 mL under present conditions; besides, you need 2 L of solution, not 100 mL.

So you do a little mental division and then multiplication.

- First – the volume. You need to prepare 2 L or 2,000 mL of solution. That is, 2,000 mL/100 mL = 20 times larger than the 100 mL volume. Remember the factor 20.
- Next, the mass of NaCl. The logic you use here is simple: if the volume of the solvent is 20 times larger then the mass of the solute must be 20 times larger too. Concentration, any concentration is a *ratio* and we have to maintain this ratio. So you need 0.9 [g] × 20 = 18 g of NaCl. Of course you cannot weigh anything in your tent but the NaCl tablets are exactly 1 g (meaning "gram") each.

So you crack the seal and open one bottle of water and put nine tablets of NaCl in it and screw the cap back on; you do the same with another bottle. (As you are shaking the bottles to dissolve the NaCl tablets you make a little mental note to contact the NaCl supplier and notify them that 1 Gm means one gigameter, that is, a billion of meters while the symbol for gram is g.) The next minute you walk out of your tent and hand the paramedic two bottles of freshly prepared saline isotonic solution; mission accomplished.

A note on concentrations: Percent concentrations, like the saline solution, are usually given as a number of grams of solid solute dissolved in a 100 mL of solution. They are good enough where high accuracy is not needed. A closer look at percentage concentration shows that this is not a very precise way of giving the ratio of solute to solvent; the densities and volumes of solute differ and the solvents change volume with temperature. The concentrations chemists and biochemists prefer to use are *molarity*, M, expressed as a number of moles of solute per 1 L of solution and *molality*, m, expressed as a number of moles of solute per 1 kg of solvent. Since the volume of many liquids changes with temperature, causing a change in molarity of a solution, it may seem that molality is a physically more correct description of a solution. In fact the opposite is true: the two concentrations are based on different physical contributions, giving an advantage to molarity at higher concentrations [1]. Physical chemists prefer molar fractions. Preparing a solution of accurate concentration can be a tedious task but understanding concentrations and knowing how to prepare a solution is important – for it may save someone's life. The next set of exercises will thoroughly refresh your knowledge of three types of concentrations.

Problem 5.2 | A moment in the life of a chemical prep-room assistant.

It is your first day as a chemical laboratory prep-room assistant. You are given a notebook sheet with instructions to prepare two NaOH solutions, one in water and another in ethanol. First, of course, you will wear gloves and put goggles on your eyes; you may also wear a lab coat or a plastic apron if available. For solution (A), you will weigh 17.323 g of NaOH and dissolve it in 500.00 mL of water at room temperature (density, $d_{HOH} = 997.41$ kg m^{-3}). For solution (B) you have to dissolve (it takes a long time) 0.054 g of NaOH in 250.00 mL of EtOH ($d_{EtOH} = 0.790$ g cm^{-3}). What are the molarity and molality of solution (A) and of solution (B)? What is the molar fraction of water in (A)?

» Solution A – Strategy and Calculation

Let us start with molarity, M. It is defined as a number of moles of solute in 1 L (or 1,000 mL) of solution. Note – *solution*, not solvent. Let us first find the number of moles of the solute NaOH in 17.323 g; we will divide this mass by the molar mass (also cited as formula weight in some literature sources):

$$n(NaOH) = 17.323\,[g]/39.997\,[g\,mol^{-1}] = 0.433\,[mol]$$

Now – think – when you add more than 17 g of NaOH to 500.0 mL of water the volume will increase. It will increase by the volume taken by 17.323 g of sodium hydroxide. And how much is that? You will find this from the relation between the mass of NaOH and its density; densities are usually given in g cm^{-3}. Insert the correct value for the density of NaOH and you should get the volume of the dry NaOH:

$$\text{Volume} = \text{mass/density}$$

Or, using the proper numbers and units we get

$$V(NaOH) = 17.323\,[g]/2.13\,[g\,cm^{-3}] = 8.133\,cm^{-3}$$

Let us pause now – I need to make a couple of assumptions and comments here.

» Assumptions #1 and 2

- We assume that 8.133 cm^{-3} of dry NaOH is very close to 8.133 mL.
- In this problem we will ignore the so-called *partial molar properties*. (They will be fully addressed in the next problem.)

When you add 17.323 g of NaOH to 500.0 mL of water you will have approximately 508.13 mL of NaOH solution. Now it is easy to calculate the molarity of this solution: it equals the number of moles of NaOH divided by the number of liters of the solution:

$$c_M = 0.433\,[\text{mol}]/0.5081\,[\text{L}] = 0.852\,\text{M} \tag{5-1}$$

The volume of the solution can change and does change with temperature – as we know the higher the temperature the larger the volume – so physical chemists decided to introduce a more accurate measure of concentration which does not depend on the temperature, molality m. Once again, it is defined as the number of moles of solute divided by the mass of the solvent in kilograms. You know how many moles of NaOH there are in 17.323 g. Now you will have to figure out the correct mass of the 500 mL of water at room temperature. Yes, it is close to 500 g but not quite the same; multiply the volume of water by its density:

$$m(H_2O) = \text{volume} \times \text{density} = 0.500\,[\text{L}] \times 0.997\,[\text{kg L}^{-1}] = 4.987 \times 10^{-1}\,\text{kg}$$

Molality is calculated as the number of moles of solute per kilogram of solvent:

$$c_m = 0.433\,[\text{mol}]/4.987 \times 10^{-1}\,\text{kg} = 0.869\,\text{m} \tag{5-2}$$

You see, molality and molarity are close but not the same. To calculate the molar fraction of water we need to know the number of moles of both components in the solution, $n(\text{NaOH})$ and $n(H_2O)$. You have already calculated the number of moles of sodium hydroxide; for water, you should divide the given mass of water by its molar mass (oftentimes called, wrongly, *formula weight*), according to the following expression:

$$n_{H_2O} = 4.987 \times 10^{-1}\,\text{kg}/18.02 \times 10^{-3}\,[\text{kg mol}^{-1}] = 27.67\,\text{mol}$$

From there we get for the total number of moles of all components in the solution – water and sodium hydroxide:

$$n_{\text{total}} = n_{\text{NaOH}} + n_{H_2O} = 28.104\,\text{mol}$$

$$x_{H_2O} = n_{H_2O}/n_{\text{total}} = 27.67/28.10 = 0.9846$$

Note: molar fractions are ratios and therefore have no units.

≫ Solution B – Calculation

You should use pretty much the same procedure as in (A): find the number of moles of solute and solvent, then the total volume of the solution, and calculate the molarity and molality. The number of moles of the solute, $n(\text{NaOH})$, is obtained when you divide the mass of NaOH by its molar mass:

$$n(\text{NaOH}) = 0.054\,[\text{g}]/39.997\,[\text{g mol}^{-1}] = 1.350 \times 10^{-3}\,\text{mol}$$

The *volume* of the dry NaOH, before you add it to the solution, will be so small that you can ignore it (check for yourself if you do not believe me). So the molarity, M, will be

$$c_M = 1.350 \times 10^{-3}\,[\text{mol}]/0.250[\text{L}] = 5.400 \times 10^{-3}\,\text{M}$$

And the molality, given as the number of moles per kilogram of the solvent, is (do not forget to convert d_{EtOH} from g cm^{-3} to kg L^{-1})

$$c_m = 1.350 \times 10^{-3}[\text{mol}]/0.250[\text{L}] \times 0.790[\text{kg L}^{-1}] = 6.836 \times 10^{-3} \, m$$

Just as I told you – the molarity and the molality of the same solution could be quite different.

A note on partial molar properties: In case you are beginning to wonder why there are so many questions and problems about concentrations I will answer by telling you that you need concentrations in about four out of every five problems in physical chemistry. The matter of fact is that a lot of chemistry and all of biochemistry takes place in solutions. Then there are problems inherent to solutions. Solutions are considered simple physical mixtures of two or more different kinds of molecules, with no chemical bonds made or broken. For a really "well-behaved" solution physical chemists have a name, by analogy with the gas laws: an *ideal solution*. Yet solutions are actually complicated systems whose molecular nature we are only now beginning to understand [1, 2, 3, 4]. Two solvents, when mixed, often release heat (or absorb heat) and undergo change in volume. Think of a water:sulfuric acid (*caution!*) mixture or a water:DMSO (dimethyl sulfoxide) mixture. After the solvent mixture equilibrates you will find that its volume is not equal to the sum of the volumes of the pure solvents (it is usually smaller). In physical chemistry we treat these problems by using the concept of molar volume, V_m. Molar volumes are empirical numbers – they are determined by experimental measurements for different solvent compositions. Read the next problem.

Problem 5.3	**More horrors from the chemistry prep-room.**

A stockroom assistant is preparing 1 L of 0.02 M ethanol/water solution of sodium hydroxide. He weighs 0.80 g of NaOH ($m_m = 40.00$ g mol^{-1}) and dissolves it in 200.00 mL distilled water ($m_m = 18.02$ g mol^{-1}, $d = 1.00$ g mL^{-1}). After filling the volumetric flask with neat ethanol ($m_m = 46.06$, $d = 0.79$ g cm^{-3}) up to the 1,000.00 mL mark, he leaves the solution on the counter and tends to another task. After some time, the components of the solution completely mix and reach their equilibrium states. At this particular ratio, 80:20, the experimental molar volumes are given as $V_m(H_2O) = 16.80$ mL mol^{-1} and $V_m(EtOH) = 57.40$ mL mol^{-1}. So when he returns, the stockroom assistant finds to his surprise that the level of the solution in the volumetric flask has dropped significantly below the 1,000.00 mL mark. Without much thinking, he adds ethanol up to the mark. (A) How much ethanol did he have to add to reach the 1,000.00 mL? (B) Was he right to add ethanol to the solution to make up to 1,000.0 mL?

≫ Solution – Strategy and Calculation

Note the words "ethanol/water solution," and "level drop." This exercise describes the typical behavior of a mixture of solvents. Probably the best approach to this type of problems is to figure out the *number of moles* of all components in the mixture. The properties of the mixture and its components may change but the number of moles will remain constant. For water, we get the number of moles by dividing the total mass of water by its molar mass:

$$n_{H_2O} = 200.00\,g/18.02\,g\,mol^{-1} = 11.10\,mol$$

You use the same procedure for ethanol and you get

$$n_{EtOH} = 800.00\,mL \times 0.79\,g\,mL^{-1}/46.04\,g\,mol^{-1} = 13.72\,mol$$

Now we mix water and ethanol and let it sit on a bench. After a while, about an hour, the *partial molar volumes* of water and ethanol have changed and the total volume is no more 1,000.0 mL. To find out the total volume of water you will multiply the number of moles of water – unchanged in pure water and in mixture – by the molar volume of water in a 2:8 mixture with ethanol. This will give you the following relation:

$$V_{H_2O} = n_{H_2O} \times V_m\,(H_2O) = 11.10\,mol \times 16.80\,mL\,mol^{-1} = 186.5\,mL$$

As you can see, this is not 200.00 mL anymore. Use the same procedure for ethanol and you will get

$$V_{EtOH} = 13.72\,mol \times 57.40\,mL\,mol^{-1} = 787.5\,mL$$

Again – less than the 800.00 mL we had at the beginning. The total volume of the solvent mixture is the sum of the two volumes:

$$V_{total} = 186.5\,mL + 787.5\,mL = 974.0\,mL$$

So, the water: ethanol mixture is 1,000.0 mL – 974.0 mL = 26.0 mL short of full liter and this is the answer (A). Note also that we have ignored NaOH in the solution and its molar properties.

(B) The question now is what to do next. The prep-room assistant took a bottle of pure ethanol and filled the volumetric flask up to the 1,000.0 mL mark. Wrong! – By doing this he has changed the composition of the solvent mixture. One should instead prepare 30–40 mL of the same ethanol:water mixture (80:20 v/v), let it equilibrate, and use it to refill the 1 L volumetric flask up to the mark.

Let us move now to more exciting things

References

Concentrations/Solutions/Liquids

1. Rösgen J, Pettitt BM, Bolen, DW (2004) Uncovering the basis for nonideal behavior of biological molecules. Biochemistry, 43:14472–14484
2. Zwier, T (2009) Squeezing the water out of HCl(aq). Science, 324:1522–1523 (perspectives)
3. Tolmakoff A (2007) Shining light on the rapidly evolving structures of water. Science 317: 54–55 (perspectives)
4. Hill TL (1957) Theory of solutions. I. J Am Chem Soc 79:4885–4890

6 Chemical Reactions and Gibbs Free Energy

We have shown that when two things are mixed together we can calculate the entropy of mixing. We can also calculate the concentration of this mixture, as we did in the previous three examples. A most complete description of the chemicals and their mixtures is by using *Gibbs free energy* (we say it is free because it does not contain any pV work, like when we are working with gases). Let us think for a moment of the following mixture: a solution of urea, $O=C(NH_2)_2$, in water, H_2O. Water, a liquid, is present in larger quantity so we call it the solvent. Urea, a solid, is present in smaller quantity and is called the solute. Before we add urea to water and completely dissolve it urea is a pure chemical. If we determined (through an experiment carried out at standard pressure, $p = 101{,}325$ Pa, and temperature, $T = 298.15$ K) its Gibbs free energy we would call it the *standard Gibbs energy*, $G_T^\ominus$, and write

$$G(\text{urea}) = G_T^\ominus(\text{urea})$$

Every material thing has a standard Gibbs free energy and we can write a similar equation for water:

$$G(H_2O) = G_T^\ominus(H_2O)$$

But when we mix urea and water and completely dissolve it, urea is still there, only somewhat changed; its Gibbs energy has changed too [1] and we write

$$G(\text{urea, mixture}) = G_T^\ominus(\text{urea}) - RT \ln a\,(\text{urea, mixture}) \tag{6-1}$$

The "correction" to the standard Gibbs energy, $G_T^\ominus$, contains the gas constant (used to connect different units), the temperature, and a logarithm of "a." We call "a" the *activity*; it is telling us how much urea, now mixed with water (and invisible to the naked eye), is *active* in this mixture. We often do not know or cannot easily measure the activity so we approximate it by concentration:

$$G \approx G_T^\ominus - RT \ln c \tag{6-2}$$

This is usually a good approximation when the concentration is low, like 10^{-3} M or lower.

P-P. Ilich, *Selected Problems in Physical Chemistry*,
DOI 10.1007/978-3-642-04327-7_6, © Springer-Verlag Berlin Heidelberg 2010

Make a note: *Gibbs energy, standard Gibbs energy, activity.*

A note on activities: When you mix water, (A) in Fig. 6.1, with urea (B), you get a mixture (C). The urea molecules are now surrounded by the water molecules and the water molecules are surrounded by the urea molecules. If the properties of urea have not changed, or have not changed much (say, less than 1%), this implies that the properties of water have not changed much either. We say that this mixture of water and urea is an *ideal solution*.

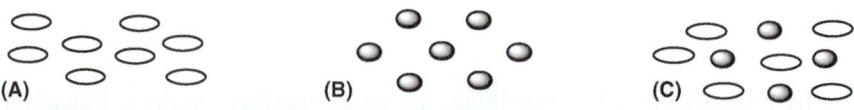

■ **Fig. 6.1** A scheme of pure solvent (**A**), a solute (**B**), and a well-behaved solution (**C**)

Very dilute solutions behave like ideal solution; however, the fact is that all solutions are *real solutions*. In a real solution the solute and the solvent do interact with each other and change something of each other's nature; typically, solute breaks the structure of solvent molecules, as schematized in Fig. 6.2(A). When talking about solvent structure we do not mean a rigid structure, like in solid bodies; it is more a way in which solvent molecules prefer to lump together. This is the case with the solutions where solute is in higher concentration. However, concentration is oftentimes not a good measure of the properties of either solute or solvent in the mixture. There are several reasons for that but two major ones can be depicted by simple schemes. In the case depicted in Fig. 6.2(B) the solute molecules prefer to interact with each other and stick together (or self-aggregate); they make smaller or bigger lumps and their *effective concentration* is *smaller* than their actual concentration. In the case shown in Fig. 6.2(C) the solvent molecules prefer to interact with solute molecules and make clusters of what we call solvent-dressed solute molecules. The solute molecules appear bigger than they are and a significant portion of solvent molecules is taken up by these solute–solvent clusters. In this case the effective concentration of the solute is *larger* than its actual concentration.

■ **Fig. 6.2** Schemes depicting a change in the solvent organization, caused by the presence of solute (**A**), self-aggregation of the solute (**B**), and strong solute–solvent aggregation (**C**)

Effective concentration corresponds to – but is not quite the same as – activity, a. We express activity as a product of concentration and activity coefficient; for molar concentrations we write

$$a = \gamma_c \times c_M \qquad (6\text{-}3)$$

When a mixture behaves almost like an ideal solution we say that the activity coefficient is close to 1 and that activity and concentration are the same. As a rule, we assume that for diluted solutions ($10^{-4}-10^{-6}$ M) the activity coefficient is sufficiently close to one. However, for many solutions, particularly the bodily fluids – blood, sweat, tears, saliva, gastric juices – the activity coefficients are quite different from 1 (smaller or bigger) [2]. So in order to truly understand the chemistry of life we have to pay close attention to activities and activity coefficients.

» Example – Δ G and chemical reactions
When sodium hydroxide, NaOH, and hydrogen chloride, HCl, are mixed they undergo a chemical reaction. In the reaction the bond between sodium and hydroxide in Na–OH is broken to give Na and OH. Hydrogen chloride breaks to give H and Cl. This is a simple reaction and you probably know what will be its products: NaCl and HOH; salt and water. But then, there is no reason why sodium, Na, and hydrogen, H, cannot form a chemical bond to give sodium hydride, NaH. By the same argument hydroxide, HO, and chlorine, Cl, would give HOCl, hypochlorous acid. So two reactions can happen:

$$Na-OH + H-Cl = NaCl + HOH \qquad (A)$$

$$Na-OH + H-Cl = NaH + HOCl \qquad (B)$$

Which way is the reaction between NaOH and HCl going to go, (A) or (B)?

» Solution
One way to figure this out is to let this reaction happen and then analyze the products; this could take time, requires a lot of skill and also access to analytical instruments. Another way, without a need for carrying out an actual reaction, is to figure out the change in G for reaction (A) and the change in G for reaction (B) and compare the two numbers. So for reaction (A), the change in Gibbs free energy will be given by

$$\Delta G(A) = G(NaCl) + G(HOH) - G(NaOH) - G(HCl)$$

And for (B) we write

$$\Delta G(B) = G(NaH) + G(HOCl) - G(NaOH) - G(HCl)$$

We calculate ΔG of these reactions by subtracting the Gibbs free energies of the *reactants* from the Gibbs free energies of the *products*; we always do it this way. After looking through thermodynamic tables – or using a computer and molecular modeling software to calculate the total energy – you will obtain the following numbers (they may differ a little, depending on the source):

$$NaOH + HCl = NaCl + HOH \qquad \Delta G\,(A) = -420\,kJ\,mol^{-1}$$

$$NaOH + HCl = NaH + HOCl \qquad \Delta G\,(B) = +264\,kJ\,mol^{-1}$$

These numbers will look better if we display them as a scheme, Fig. 6.3. The graph tells us that, when NaOH and HCl react to give NaCl and HOH, the total Gibbs free energy will drop by 420 kJ for each mole of reactants and products. We mentioned before (in the problem of skating on ice and water) the important convention about the sign of the total energy. The same rule applies to the change of energy: if the total energy decreases this reaction is *spontaneous*. Physical chemists call this an *exergonic* reaction.

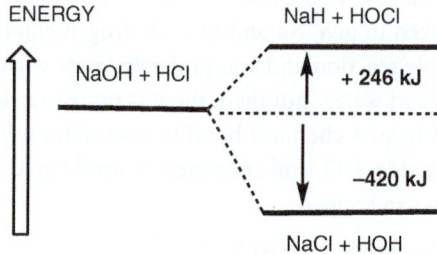

ENERGY

NaOH + HCl   NaH + HOCl   + 246 kJ

−420 kJ

NaCl + HOH

■ **Fig. 6.3** The energy budget of two NaOH + HCl reactions

It is the opposite for reaction (B): we have to add, through heating, light, or other agency, 264 kJ of energy in order to make it happen. We say that reaction (B) has to be *driven* or that the reaction is *endergonic*. So sodium hydroxide and hydrogen chloride can give sodium hydride and hypochlorous acid, only the reaction is expensive in terms of energy; we have to pay for it. So it just happens that by mixing NaOH and HCl (*careful!*), you will more likely get water and kitchen salt (and heat!).

References

Concentrations/Solutions/Liquids

1. Hill TL (1957) Theory of solutions. I. J Am Chem Soc 79:4885–4890
2. Rösgen J, Pettitt BM, Bolen DW (2004) Uncovering the basis for nonideal behavior of biological molecules. Biochemistry, 43:14472–14484

7 Gibbs Free Energy and Chemical Equilibria

I already said that it is important to understand and to know how to calculate and prepare different concentrations. Many chemical reactions – and *all* the biochemical reactions that run our bodies – take place in a solution. Now we are going to put together a formula for the Gibbs energy of a chemical in solution and a formula for calculating the Gibbs energy in a chemical reaction. For the reaction given above we write

$$\Delta G(A) = G(NaCl) + G(HOH) - G(NaOH) - G(HCl)$$

Now we replace each $G(NaCl)$, $G(HOH)$, $G(NaOH)$, $G(HCl)$ with the expression for the Gibbs energy of a substance in a mixture:

$$G = G_T^{\ominus} + RT \ln a$$

$$\Delta G(A) = (G_T^{\ominus} + RT \ln a)(NaCl) + (G_T^{\ominus} + RT \ln a)(HOH)$$
$$- (G_T^{\ominus} + RT \ln a)(NaOH) - (G_T^{\ominus} + RT \ln a)(HCl) \tag{7-1}$$

When we sort out all the terms and make some simplifications we get

$$\Delta G = \Delta G_T^{\ominus} + RT \times \{\ln a(NaCl) + \ln a(HOH) - \ln a(NaOH) - \ln a(HCl)\}$$

The first term, $\Delta G_T^{\ominus}$, stands for the sums and differences of all the $G_T^{\ominus}$ values in this reaction. Since you don't know the activities you may try concentrations; if they are low, the equation will hold. We then write

$$\Delta G = \Delta G_T^{\ominus} + RT \times \ln\{[NaCl][HCl]/[NaOH][HOH]\} \tag{7-2}$$

Here we used two properties of logarithms, one is given as $\ln x + \ln y = \ln (xy)$ and the other is given as $\ln x - \ln y = \ln (x/y)$. The ratio of concentrations of the products and reactants is called *reaction quotient* and labeled by Q:

$$Q = \{[NaCl][HOH]/[NaOH][HCl]\} \tag{7-3}$$

Now you can write a simpler-looking equation, where $\Delta G_T^{\ominus}$ is the sum and difference of individual $G_T^{\ominus}$, as in (7-1):

P.-P. Ilich, *Selected Problems in Physical Chemistry*,
DOI 10.1007/978-3-642-04327-7_7, © Springer-Verlag Berlin Heidelberg 2010

$$\Delta G = \Delta G_T^{\ominus} + RT \times \ln Q \tag{7-4}$$

At the beginning of the reaction, before we mix the solutions of NaOH and HCl, there is no change in the total G. As the reactants come to contact they start reacting, yielding water and sodium chloride and releasing *heat*. How much heat? The graph, Fig. 6.3, tells us there is about 420 kJ of heat released for each mole of NaOH and HCl. The reaction Gibbs energy becomes negative, $\Delta G < 0$, as shown in the graph. After some time, when NaOH and HCl are used up, the reaction Gibbs free energy will reach a *minimum* and will not change anymore. So the Gibbs free energy at this point in reaction is –420 kJ mol^{-1}, but the *change* in Gibbs energy is zero: $\Delta G = 0$. We say that the reaction has reached *equilibrium* and the quotient Q is a ratio of equilibrium concentrations (or, more accurately, activities) of products and reactants. Such quotient is a constant for each chemical reaction and we have a name for it: we call it an *equilibrium constant*, K_{eq}. So for a chemical reaction at equilibrium we write

$$\Delta G = \Delta G_T^{\ominus} + RT \times \ln K_{eq} \tag{7-5}$$

Since $\Delta G = 0$ we will not mention it anymore; we move $\Delta G_T^{\ominus}$ to the left side, multiply the equation through by -1 and write

$$\Delta G_T^{\ominus} = -RT \times \ln K_{eq} \tag{7-6}$$

Equations (7-4) and (7-6) are probably *the most important equations in physical chemistry*. They connect the thermodynamic function G with a chemical reaction. The chemical reaction, after reaching equilibrium, is expressed by a ratio of the amounts of products and reactants near or at equilibrium. The amounts of products and reactants are usually expressed as concentrations, pressures (if gases are reacting), or, generally, *activities*. Since G contains both the reaction heat (enthalpy, H) and the measure of disorder at a given temperature (entropy, S) the equation above allows us to relate the concentrations of reactants and products with the reaction enthalpy and entropy and the temperature at which the reaction occurs. It helps us to better understand the energy budget of any chemical reaction – which is one of the main goals of physical chemistry.

We will start with a simple chemical reaction to learn a few tricks and then tackle problems of biological and biophysical importance, as follows: the DNA unfolding and refolding, a drug-receptor binding, an amino acid acid–base reaction, and a bio-oxidation electrochemical reaction.

| Problem 7.1 | How the old malt is actually made (in case you did not know). |

An important industrial chemical reaction is hydration of ethene to ethanol; the reaction is carried out at high pressure and the reactants and products are all in gas (vapor) phase:

$$H_2C = CH_2(g) + H_2O(g) \rightarrow CH_3CH_2OH(g)$$

At $t = 300°C$ and $p = 200$ atm, the equilibrium quotient of pressure (activities), $K_p = 0.860$. (A) Calculate the partial pressure of the product, ethanol, under these conditions. (B) In the second part, assume that K_p at half this temperature is 6.8×10^{-2} and estimate the $\Delta S^{\ominus}$ for this process.

» **Solution A – Strategy**

This riddle is telling us that two gases, ethane and water vapor, are mixed and ethanol is produced. No concentrations, activities, or partial pressures of the reactants or the product are given. We only know the quotient of their pressures:

$$K_p = p(CH_3CH_2OH)/p(H_2C = CH_2) \times p(HOH) = 0.860$$

Note: The equilibrium constant is dimensionless. The given temperature and total pressure tell us about the reaction conditions but are not useful in finding the partial pressure of the product. There are quite a few questions like this which, on first reading, seem to be lacking data and information needed to find the answer. In fact, these problems are simpler than they seem – and this is often the case in physical chemistry. Remember *little big trick* #3? When no mass, number of moles, volume, or partial pressure (for gases) has been given we assume *one*; one mole, one gram, or one atmosphere (for gases), respectively.

Now we are going to apply the very powerful *little big trick* #4: When no amount (mass, or number of moles, or concentration) of product is given you express it as a mathematical *variable*, x, and use algebra to find its value.

The beginning of the reaction, in moles of the reactants and product, is given as

$$1 \times C_2H_4(g) + 1 \times H_2O(g) = 0 \times CH_3CH_2OH(g)$$

You have zero for ethanol as there is no product at the start of the reaction.

After a while a part of ethene, and the same part of H_2O, gets converted to ethanol. This part could be, for example, 5% or it could be 95%; we do not know what it is so we write x; x moles. If all of ethene and water have reacted none of the reactants would be left and there would be 100% or 1 mol of ethanol. So x is a number between 0 and 1. The reaction equation now reads

$$(1 - x)C_2H_4 + (1 - x)H_2O = xCH_3CH_2OH \tag{7-7}$$

Let us assume that the reaction has reached equilibrium. No more of ethene and water are used up and no more of ethanol is produced, so you can write for the equilibrium constant, K_p (the subscript p is for a constant expressed in pressures)

$$K_p = x/(1-x) \times (1-x) = x/(x^2 - 2x + 1)$$

The rest of the problem is an exercise in high school math: solving a quadratic equation. When we multiply K_p with $(x^2 - 2x +1)$ and move the "x" on the right side of the equation to the left side we get

$$K_p x^2 - (2K_p + 1)x + K_p = 0$$

Now divide the equation through by $K_p = 0.86$ and you get a simplified quadratic equation $x^2 + Bx + C = 0$

$$x^2 - 3.1628 \times x + 1 = 0$$

with the following coefficients, A = 1, B = −3.1268, and C = 1.

Since this is a *second-order* algebraic equation it has *two* numerical solutions:

$$x_{1,2} = -B \pm \sqrt{[B^2 - 4A\,C]}/2A$$

$$x_{1,2} = \{3.1628 \pm \sqrt{[-3.1628^2 - 4]}\}/2$$

$$x_{1,2} = (3.1628 \pm 2.4502)/2$$

$$x_1 = 0.3563$$

Note: The second solution, $x_2 = \{3.1628 + 2.4502\}/2 = 2.8065$, is larger than the total number of moles in the reaction (which is between 1 and 2) so we will disregard it. Next, we should answer the following question: What is the total number of moles – reactants and products – in this reaction? You will find this when you insert the value for x_1 in (7-7) given above and solve it for n(total), the total number of moles of all chemicals in the reaction:

$$(1-x)C_2H_2 + (1-x)H_2O = xCH_3CH_2OH$$

$$n(\text{total}) = (1 - 0.3563) + (1 - 0.3563) + 0.3563$$

$$n(\text{total}) = 2 - 0.3563 = 1.6437$$

And the *molar fraction* of ethanol at this stage of the reaction is

$$x(\text{ethanol}) = 0.3563/1.6437 = 0.2168$$

And now it is easy to find the answer to the first question: What is the *partial pressure* of ethanol vapors in this reaction? Like any partial pressure – remember Problem 2.4 about inhaling and exhaling? – it equals the total pressure multiplied by the molar fraction of the gas (vapor). So the partial pressure of ethanol vapors in the reaction mixture is

$$p_{EtOH} = x_{EtoH} \times p_{tot} = 0.2168 \times 200[\text{atm}] = 43.36 \,\text{atm}$$

You could convert the pressure to newton/m^2 (pascal) but atmosphere is used in all of gas industry.

» Solution B – Strategy and Calculation

Let us read the question again:

(B) In the second part, assume that K_p at half this temperature is 6.8×10^{-2} and calculate the $\Delta S^{\ominus}$ for this process.

This type of question often shows up in calculations involving $\Delta G^{\ominus}$, the reaction standard Gibbs energy, and K_p, the equilibrium constant. Your key to finding the answer is in $\Delta G^{\ominus}$:

You should first convert the two temperatures to the thermodynamics scale using the following expressions:

$$T_1 = 300 + 273 = 573\,K \quad \text{and} \quad T_2 = 300/2 + 273 = 423\,K$$

Then calculate the two $\Delta G_T^{\ominus}$ using the two temperatures and the two equilibrium constants:

$$\Delta G_{573}^{\ominus} = -R \times T_1 \times \ln 0.860 = -8.314\,[\text{J K}^{-1}\,\text{mol}^{-1}] \times 573[K] \times -1.51 \times 10^{-1}$$
$$= +7.18 \times 10^2\,\text{J}$$

$$\Delta G_{423}^{\ominus} = -R \times T_2 \times \ln 6.80 \times 10^{-2}$$
$$= 8.314\,[\text{J K}^{-1}\,\text{mol}^{-1}] \times 423[K] \times -2.69$$
$$= +9.45 \times 10^3\,\text{J}$$

Now let us see what we can do about the standard entropy of this reaction; you should re-write the $\Delta G_T^{\ominus}$ in terms of $\Delta H_T^{\ominus}$ and T times $\Delta S_T^{\ominus}$:

$$\Delta G_{573}^{\ominus} = \Delta H_{573}^{\ominus} - 573 \Delta S_{573}^{\ominus} \quad \text{(A)}$$
$$\Delta G_{423}^{\ominus} = \Delta H_{423}^{\ominus} - 423 \Delta S_{423}^{\ominus} \quad \text{(B)}$$

Let us make a list of the things we know, then a list of the things we need to know, and finally a list of the things we are not quite sure what to do about.

List (1) – the things we know:

- $\Delta G_{573}^{\ominus}, \Delta G_{423}^{\ominus}, T_1$, and T_2

List (2) – the things we need to figure out:

- $\Delta S_{573}^{\ominus}$ and $\Delta S_{423}^{\ominus}$

List (3) – the things we are not sure what to do about:

- $\Delta H_{573}^{\ominus}$ and $\Delta H_{423}^{\ominus}$

Now you should apply *little big trick* #5: You reshuffle or slightly re-scale, by multiplication, for example, (A) and (B), and then subtract them to eliminate the same variables. The trick is based on the following approximations:

» Approximations
- When temperature changes $\Delta H_T^{\ominus}$ remains nearly constant.
- When temperature changes $\Delta S_T^{\ominus}$ remains nearly constant.

For most reactions these approximations hold fairly well within a small temperature range. You may want to remember this approximation as you will need it in other problems. Keep in mind the following important point: these are the *standard* enthalpy and entropy; the actual enthalpy and entropy do change with temperature.

The next step is to *subtract* (A) and (B); it does not really matter which one goes first:

$$\Delta G_{573}^{\ominus} - \Delta G_{423}^{\ominus} = \Delta H_{573}^{\ominus} - T_1 \times \Delta S_{573}^{\ominus} - (\Delta H_{423}^{\ominus} - T_2 \times \Delta S_{423}^{\ominus})$$

Now you use the first approximation, that the standard enthalpy does not change with temperature, and write $\Delta H_{573}^{\ominus} \approx \Delta H_{423}^{\ominus} = \Delta H^{\ominus}$. The second approximation allows you to write the following relation: $\Delta S_{573}^{\ominus} \approx \Delta S_{423}^{\ominus} = \Delta S^{\ominus}$. Now use these relations and rewrite the equation given above:

$$\Delta G_{573}^{\ominus} - \Delta G_{423}^{\ominus} = \Delta H^{\ominus} - T_1 \times \Delta S^{\ominus} - \Delta H^{\ominus} + T_2 \times \Delta S^{\ominus}$$

Cancel the $\Delta H^{\ominus}$ and put together the entropy terms and you get

$$\Delta G_{573}^{\ominus} - \Delta G_{423}^{\ominus} = \Delta S^{\ominus} \times (T_2 - T_1)$$

Insert the numbers for the Gibbs energies and temperatures, punch few keys on your calculator, and you will get

$$7.18 \times 10^2 \, [\text{J}] - 9.45 \times 10^3 \, [\text{J}] = \Delta S^{\ominus} \times (423 - 573) \, [\text{K}]$$

And the $\Delta S^{\ominus}$ for this reaction reads

$$\Delta S^{\ominus} = +58.2 \, \text{J K}^{-1}$$

The positive $\Delta S^{\ominus}$ change reflects in part the highly endergonic nature of the reaction – the high temperature and pressure needed to make it happen. The change in the *reaction* entropy, ΔS, is expected to be less positive, based on the fact that two reactant molecules, ethene and water, give one product molecule, ethanol, so the overall disorder of the system decreases.

A comment: This was a long way to calculate one number. You will also find out that this is a useful method as you will be able to apply it to a number of seemingly different problems and questions involving Gibbs standard energies and equilibrium constants. Note also that, once you have figured out $\Delta S^{\ominus}$, you can plug it back and calculate the change in the standard enthalpy of the reaction, $\Delta H^{\ominus}$. Time for a break, don't you think?

7.1 Receptor and Ligand Equilibria
A large number of biological reactions involve binding of one molecule to another. One of them is usually small and we call it a *ligand, L*. The other is usually large and

is called a *receptor*, R. You may think of a ligand as coming to a receptor and binding to it to form a receptor–ligand complex. We describe this process by this expression: $L + R = R{:}L$. Also, many receptor molecules can bind, and hold, more than one ligand. Binding or *association* of a ligand to a receptor does not have to involve covalent chemical bonds. Once bound, ligand and receptor can separate or *dissociate* and we say that a ligand–receptor binding is *reversible*. When the ligand association and dissociation reach equilibrium we can describe this process by an equilibrium constant. From the equilibrium constant, we can calculate the standard Gibbs free energy, enthalpy, and entropy for the receptor–ligand interaction. A knowledge of thermodynamic parameters for ligand–receptor interactions is essential for understanding the biochemical processes in our body.

Problem 7.2 | Big mamma's hug.

When you are under stress a pea-sized bulbous tissue above your kidney, called adrenal gland, secretes a chemical called cortisol. Binding of a cortisol molecule, a *ligand*, L, to glucocorticoid *receptor*, R, sends your body a signal to speed up metabolism. Experimental measurements of the association of cortisol, L, with glucocorticoid, R, to form a receptor–ligand complex, R:L, gave –8.8 kJ mol^{-1} for $\Delta H^{\ominus}$ and 121 J K^{-1} mol^{-1} for $\Delta S^{\ominus}$, at 0°C [1]. From these data calculate the association, K_A, and dissociation, K_D, constants for the cortisol–glucocorticoid interaction.

» Solution – Strategy

There are new and perhaps unfamiliar words in the riddle given above. You will do well to learn how to ignore them and look for the essence of the question here. And the essence is *the binding of a ligand to a receptor to give a receptor–ligand complex*. Translated into symbols these English words look like

$$R + L = R{:}L \tag{7-8}$$

Given are $\Delta H^{\ominus}$ and $\Delta S^{\ominus}$ and the temperature; find the K_{eq} for this "reaction". In your textbook, under the chapter "Second Law of Thermodynamics" you will find the *most important thermodynamic equation*:

$$\Delta G^{\ominus} = \Delta H^{\ominus} - T\Delta S^{\ominus} \tag{7-9}$$

In the chapter titled "Chemical Equilibrium" you will find what I think is the *most important physical chemistry equation*:

$$\Delta G^{\ominus} = -RT \ln K_{eq} \tag{7-10}$$

Good – now what to do with these equations? You need to know K_{eq}. You will calculate it from the $\Delta G^{\ominus}$ in (7-10). First, you will insert the numbers given for the standard enthalpy, standard entropy, and temperature into (7-9) and calculate the standard Gibbs free energy of the reaction, $\Delta G^{\ominus}$. Done.

» Solution – Calculation

First, the standard Gibbs energy of the reaction:

$$\Delta G^{\ominus} = \Delta H^{\ominus} - T\Delta S^{\ominus} = -8.8 \,[\text{kJ mol}^{-1}] - \{(0 + 273.2)[\text{K}]$$
$$\times (121 \,[\text{J K}^{-1}\,\text{mol}^{-1}])\}$$

Note two things:

- You have to convert 0°C to degrees Kelvin, K, by adding 273.2 to the Celsius temperature.
- The standard enthalpy is given in kJ, i.e., kilojoules or thousands of joules, while entropy is given in joules; it is very easy to overlook this detail and to get a completely wrong result from a good calculation. (You wonder why is it that one property is given in kJ and other just in J. Probably because physical chemists, like most of us, are uncomfortable with very big or very small numbers so they use the following abbreviations: k for thousand, M for million, and p for a very small, one trillionth.)

Insert the right numbers and punch the right keys on your calculator and you will get

$$\Delta G^{\ominus} = -8.8 - 33.0 \,\text{kJ mol}^{-1} = -41.8 \,\text{kJ mol}^{-1}$$

And for the K_{eq} you will use (7-10):

$$\Delta G^{\ominus} = -RT \ln K_{eq}$$

From here you get for the equilibrium constant

$$K_{eq} = \exp[-\Delta G^{\ominus}/RT] \qquad (7\text{-}11)$$

How did we get this? We used the fact that $\exp(x)$ and $\ln[x]$ are two mutually inverse functions (like multiplication and division); exp of course stands for the (very famous in mathematics) number e: $e = 2.7818281828\ldots$ (we don't know how far the decimals go); $\exp[x] = e^x$.

Note that whenever you have power calculations, e^x or 10^x, you should be careful to use the correct number for x as a small error in the argument (*argument* = everything inside the square bracket in (7-11)) will result in a large error in the result.

You get for the ligand–receptor equilibrium constant

$$K_{eq} = \exp[-(-41.8 \,\text{kJ mol}^{-1})/8.31 \,\text{J K}^{-1}\,\text{mol}^{-1} \times 273.2 \,\text{K}] = \exp[18.41] = 9.91 \times 10^7$$

A comment: Every time you have exp[something] or ln[something] the units within the brackets must cancel. If they don't you are in trouble; you have to go back and check every step. The K_{eq} you just calculated is the equilibrium constant for the following process:

$$R + L = R{:}L$$

In this reaction the ligand and the receptor *associate*, so this is called an association process (or *interaction*) and its equilibrium constant is an *association constant*, or K_A.

$$K_A = [R{:}L]/[R][L] \tag{7-12}$$

Opposite of it is when ligand and receptor fall apart or *dissociate*; the equilibrium constant for that process is known as *dissociation constant*, or K_D.

$$R{:}L = R + L$$

$$K_D = [R][L]/[R{:}L]$$

Now – it does not take a rocket scientist to figure out the relation between K_A and K_D; they are *inverse* to each other:

$$K_D = 1/K_A = 1/9.91 \times 10^7 = 1.01 \times 10^{-8} \tag{7-13}$$

Done. But wait – I have a question for you: What do these numbers, 9.91×10^7 for K_A and 1.01×10^{-8} for K_D, mean in this reaction? Remember – we are physical chemists, not numerical mathematicians; we care about the meaning, not the numbers.

I suggest you take another look at (7-12):

$$K_A = [R{:}L]/[R][L]$$

The value of the equilibrium constant, $K_A = 9.91 \times 10^7$, is orders of magnitude larger than in the case of the reaction of ethane a water vapor, $K_p = 0.86$ and 0.068. Whenever the equilibrium constant is smaller than one we say that this is a weak reaction. On the other hand, in case of cortisol and glucocorticoid receptor, the large ratio of the concentration of the complex $R{:}L$ to the product of the concentrations of free ligand, L, and receptor, R, tells us that there is very little free cortisol in presence of glucocorticoid receptor. We say that cortisol and glucocorticoid receptor have a very high affinity for each other.

Problem 7.3 | A sequel to big mama's hug.

Assuming the ligand concentration of 2.0×10^{-7} M [2] and the receptor concentration 4.0×10^{-5} M calculate the molar fraction of the $R{:}L$ complex at $0°C$.

≫ Solution – Strategy

Remember the reaction of ethene plus water (Problem 7.1) to give ethanol and *little big trick #4*? This is the right moment to use it again. We have a 200-fold concentration of receptor R. Since all reactants and products are in the same pot, we can ignore the volume and treat the concentrations as amounts. (Do you follow me?) This means that we can add and subtract concentrations like they are masses or volumes. Since the affinity of the glucocorticoid receptor for cortisol ligand is so high their reaction is controlled by a low ligand concentration, under normal conditions. The amount of the complex $R:L$ is zero before the reaction start but, once the reaction gets under way, it becomes larger; it becomes a positive number which we will label as x. So you write, similarly to Problem 7.1, the following relation:

$$(2.0 \times 10^{-7} - x)L + (4.0 \times 10^{-5} - x)R = xR:L$$

You get the total number of moles when you add the moles of *all* compounds present in the pot. You are counting the beans now, not doing the math, so you add all terms whether they are from the left or right side of the equation:

$$(2.0 \times 10^{-7} - x) + (4.0 \times 10^{-5} - x) + x = 4.02 \times 10^{-5} - x$$

Again, we do not need parentheses; in the previous expression I used them to neatly separate the terms. The value of x is calculated from the equilibrium equation:

$$K_A = x/(4.0 \times 10^{-5} - x)(2.0 \times 10^{-7} - x)$$

Since we have already done similar problems I am going to just outline the solution:

$$K_A = x/(8.0 \times 10^{-12} - 2.0 \times 10^{-7} \times -4.0 \times 10^{-5}x + x^2)$$

Multiply what has to be multiplied, add what has to be added, and move all terms to the left side and you should get

$$K_A x^2 - (4.02 \times 10^{-5}K_A + 1)x + 8.0 \times 10^{-12}K_A = 0$$

Now insert the value for the equilibrium constant:

$$9.91 \times 10^7 \times x^2 - (9.91 \times 10^7 \times 4.02 \times 10^{-5} + 1)x + 9.91 \times 10^7 \times 8.0 \times 10^{-12} = 0$$

Do the required multiplications and additions, then divide the equation throughout by 0.18, and you will get a much simpler expression:

$$x^2 - 4.021 \times 10^{-5}x + 8.0 \times 10^{-12} = 0$$

where $A = 1$, $B = -4.02 \times 10^{-5}$, and $C = 8.0 \times 10^{-12}$, with two solutions given as

$$x_{1,2} = -B \pm \sqrt{\{B2 - 4AC\}/2A}$$

Insert the numbers for A, B, and C and you will get

$$x_{1,2} = \{4.02 \times 10^{-5} \pm \sqrt{[(-4.02 \times 10^{-5})^2 - 4 \times 8.0 \times 10^{-12}]}\}/2$$

The two solutions to the quadratic equation are given as $x_1 = 4.00101 \times 10^{-5}$ and $x_2 = 1.99949 \times 10^{-7}$. We started this binding reaction with 4.0×10^{-5} mol of receptor and 2.0×10^{-7} mol of ligand; clearly the first solution, $x_1 = 4.00101 \times 10^{-5}$, does not make sense. Why? – Because you cannot get 4.00101×10^{-5} moles of the $R{:}L$ complex from 2.0×10^{-7} mole of L, right? The second solution, $x_2 = 1.99949 \times 10^{-7}$, looks reasonable and you will use it to calculate the total number of moles, $n(L) + n(R) + n(R{:}L)$, of the reagents which are taking part in the reaction:

$$n(tot) = (2.0 \times 10^{-7} - x) + (4.0 \times 10^{-5} - x) + x$$

$$n(tot) = (2.0 \times 10^{-7} - 1.99949 \times 10^{-7}) + 1.99949 \times 10^{-7} = 2.0005 \times 10^{-7}$$

And the molar fraction of the $R{:}L$ complex is

$$x(R{:}L) = n(R{:}L)/n(tot) = 1.99949 \times 10^{-7}/2.00005 \times 10^{-7} = 0.9995$$

This is a high number, which makes sense given that almost all of the cortisol ligands are bound to the receptor ($K_A = 9.91 \times 10^7$).

A comment: There is an interesting comparison to be made. Go back and look through the reaction of conversion of ethene to ethanol, Problem 7.1. Then, compare it with how you have analyzed the interaction of cortisol–glucocorticoid receptor association, Problem 7.3. In both cases you used pretty much the same thinking and applied the same algebra. This is yet another example of the power of physical chemical thinking. In one case you apply it to a large-scale industrial chemical reaction at high temperature and pressure. In another case you use the same procedure to a subtle interaction between a hormone and its receptor taking place in our bodies. But the way to solve the two problems was the same.

A note on multiple ligands and receptors: In the interaction of cortisol (a stress hormone)–glucocorticoid receptor we stated that one ligand can bind to one receptor. In nature, processes and interactions are usually more complicated. Receptors are large molecules and can bind and hold two, three, or more ligands. For the binding of the first ligand we write

$$L + R = R{:}L$$

and the equilibrium constant reads

$$K_A(1) = [R{:}L]/[R][L]$$

Now comes the second ligand which binds to $R{:}L$ to give $R{:}L_2$. We write

$$L + R{:}L = RL_2 \tag{7-14}$$

and the equilibrium constant is a quotient of the product(s) and reactants:

$$K_A(2) = [R{:}L_2]/[R{:}L][L] \tag{7-15}$$

And so on for the 3rd, 4th, or nth ligand. (A hemocyanin molecule, Hc – the dioxygen carrier in tarantula blood – has 24 ligand binding sites.) A good example of a receptor with more than one binding site is human hemoglobin, HbA, in our red blood cells: it has four binding sites and can bind and carry up to four O_2 molecules. How do we keep track of binding interactions, number of ligands, and equilibrium constants for such receptors?

Let us now apply this math to the HbA molecule:

(1) First O_2 molecule: $O_2 + HbA = HbA{:}O_2$ and the equilibrium constant is given as $K_A(1) = [HbA{:}O_2]/[HbA]\,pO_2$
pO_2 is the partial pressure of O_2 in the alveolar chambers in your lungs. (In a problem way back, Problem 2.4, when we were talking about *ideal gases*, you may find these pressures as 0.209 atm for inhalation and 0.153 atm for exhalation.)

(2) Second O_2 molecule:

$$O_2 + HbA{:}O_2 = HbA{:}(O_2)_2 \qquad K_A(2) = [HbA{:}(O_2)_2]/[HbA{:}O_2]pO_2$$

If you now look at the first equilibrium you may see that, after we reshuffle it a little, we can write $HbA{:}O_2$ as $K_A(1) \times [HbA] \times pO_2$. Use this substitution to re-write the expression for the equilibrium constant for binding of second O_2:

$$K_A(2) = [HbA{:}(O_2)_2]/K_A(1) \times [HbA] \times pO_2 \times pO_2$$

Or, you may write this as

$$K_A(2) \times K_A(1) = [HbA{:}(O_2)_2]/[HbA] \times (pO_2)^2$$

Now it does not take much pondering to figure out that the equilibrium constant for the fourth O_2 can be written as

$$K_A(4)K_A(3)K_A(2)K_A(1) = [HbA{:}(O_2)_4]/[HbA] \times (pO_2)^4$$

This reshuffling allows us to keep track of fewer variables, three concentrations rather than six, in the expression for the fourth O_2 ligand.

A note on cooperativity: You see that such processes – and the math used to describe them – can get rather complicated. There is more to it: after the first O_2 molecule binds to hemoglobin, to form the $HbA{:}O_2$ complex, the hemoglobin changes shape and other properties to accommodate binding of the second O_2. This is even more true in the case of third and fourth O_2 molecule; the more saturated the hemoglobin molecule is, the more likely it is to bind the next dioxygen ligand.

Imagine that you are hugging or giving shelter and protection to a creature in need of help; by doing so you will probably be more inclined to try to protect yet another such creature. When something like this occurs we say that the receptor exhibits *positive cooperativity*. Human hemoglobin exhibits positive cooperativity but many other receptors do not. Now we can construct a really interesting system that contains ligands, receptors, and, in addition, *a gate* that allows the ligands but not the receptors to pass through. There are many examples of this situation in our body, kidneys being probably the best known.

Problem 7.4 | **No entry for big guys (a gated community).**

In an artificial kidney there is a compartment filled with a solution containing ligands L and receptors R. There is a membrane in the middle, separating the compartment into the left chamber and the right chamber. The receptors and the ligands interact to form a receptor:ligand complex, $R{:}L$, with an association constant K_A. The solvent molecules, H_2O, and ligand can pass freely through the membrane in either direction. The receptor, a large molecule, and the receptor–ligand complex, $R{:}L$, cannot pass through the membrane and remain in the left chamber only. (Such membranes are called semi-permeable membranes. An example of a semi-permeable membrane is human skin: it lets perspiration leave your body but it does not allow the water from a shower or a swimming pool enter the body.) In the left chamber the total receptor concentration is 3.7200×10^{-3} M. The total ligand concentration in the same chamber is 0.5000 M. In the right chamber the total ligand concentration is 0.4995 M. (A) Use these data to calculate K_A at room temperature, $T = 298.2$ K.

» Solution A – Strategy and Calculation

This is a large problem that may appear intimidating at first sight. On the other hand – you have already found your way through several physical chemical riddles, some quite complicated, and should not get scared so easily. It will help us to make a list of the major items mentioned above:

- A semi-permeable membrane,
- Ligand–receptor interaction,
- Equilibrium constant, K_A,
- $\Delta G^{\ominus}$,
- Two temperatures,
- A constant $\Delta H^{\ominus}$, and
- $\Delta S^{\ominus}$.

We have seen all of this before except the semi-permeable membrane. Let us start with question (A): What is K_A at room temperature? We have to figure out how to calculate K_A from the concentrations given for the left chamber and the right chamber. Perhaps a scheme of this experimental setup will help us – a picture is worth 10,000 (or – was it a 1,000?) words, goes an old saying. The scheme of an artificial kidney is given in Fig. 7.1:

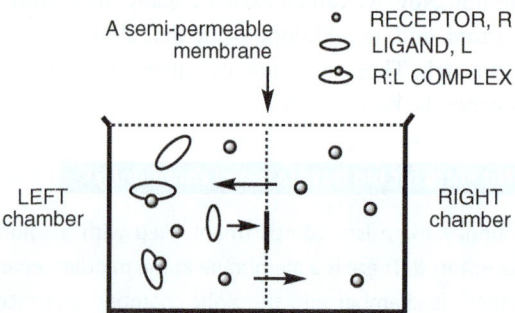

■ **Fig. 7.1** A scheme of a dialysis setup: two chambers, semi-permeable membrane, L, R, and L:R complex

Let us now check what we know about the concentrations of the molecule and complexes in this solution. The riddle says that the *total* concentration of the ligand is 0.5000 M in the left chamber and 0.4995 M in the right chamber. What does *total ligand concentration* mean? Now we have to use chemical reasoning: ligand exists as free, i.e., when it is not bound to receptor, $[R]_{free}$, and as ligand bound to receptor in the ligand–receptor complex, $[R{:}L]$. There is no third form of ligand in this solution. So the *total* ligand concentration must be a sum of the concentrations of *free* ligand and the concentration of *bound* ligand:

$$[L]_{tot} = [L]_{free} + [R{:}L] \tag{7-16}$$

A brief comment on concentrations and amounts: Concentrations are *specific* properties, defined for certain volumes, and for two solutions, they may not be added, subtracted, or multiplied. However, when all molecules, L, R, and R:L are in the same solution the volume is the same for all solutes and we can manipulate the concentrations the same way we do with masses and numbers of moles: we can add and subtract them. Make a note of this.

A look at the dialysis chamber setup, Fig. 7.1, above tells us the following:

- In the left chamber ligand exists in both forms, as free ligand and as ligand–receptor complex, and we write

  $$[L]_{tot} = [L]_{free} + [R{:}L] = 0.500 \, M$$

- In the right chamber there is only free ligand, as no receptor is allowed

  $$[L]_{tot} = [L]_{free} = 0.4995 \, M$$

Since the ligand can pass freely between the left and right chambers then, by the requirement of chemical equilibrium, the concentration of free ligand must be the same in both chambers: $[L]_{\text{free}}(right) = [L]_{\text{free}}(left) = 0.4995$ M. You may now rewrite the expression for the total concentration of ligand in the left chamber as the following:

$$[L]_{\text{tot}} = [L]_{\text{free}} + [R{:}L]$$

Next, we insert the numbers given for $[L]_{\text{tot}}$ and $[L]_{\text{free}}$:

$$0.5000 \,\text{mol L}^{-1} = 0.4995 \,\text{mol L}^{-1} + [R{:}L]$$

This gives us the concentration of the ligand–receptor complex (found only in the left chamber):

$$[R{:}L] = 0.5000 - 0.4995 = 0.0005 = 5.0 \times 10^{-4} \,\text{mol L}^{-1}$$

Let us now write down what we know about the receptor:

- Receptor is found only in the left chamber as free receptor, $[R]_{\text{free}}$, and as receptor bound with ligand in a complex, $[R{:}L]$.
- The total receptor concentration is a sum of free and bound receptors: $[R]_{\text{tot}} = [R]_{\text{free}} + [R{:}L]$
- The total receptor concentration is given: $[R]_{\text{tot}} = 3.7200 \times 10^{-3}$ M.
- Concentration of the receptor–ligand complex, $[R{:}L]$, is 5.0000×10^{-4} M, as we calculated above.

You should re-write the expression for the total receptor concentration as

$$[R]_{\text{tot}} = [R]_{\text{free}} + [R{:}L] \tag{7-17}$$

and insert the numbers we know:

$$3.7200 \times 10^{-3} \,\text{M} = [R]_{\text{free}} + 5.0000 \times 10^{-4} \,\text{M}$$

If you punch the right keys on your calculator you should have for the concentration of free receptor

$$[R]_{\text{free}} = 3.7200 \times 10^{-3} - 5.0000 \times 10^{-4} = 3.2200 \times 10^{-3} \,\text{M}$$

Now you have everything you need to calculate the association constant at $T_1 = 298.2$ K and, further down the road, $\Delta G^{\ominus}_{298}$. For the interaction $R + L = R{:}L$ the equilibrium (or association) constant is given as

$$K_{\text{eq}}^{298} = [R{:}L]/[R][L]$$

where $[R]$ and $[L]$ are $[R]_{\text{free}}$ and $[L]_{\text{free}}$, that is,

$$K_{\text{eq}}^{298} = [R{:}L]/[R]_{\text{free}} \times [L]_{\text{free}} \tag{7-18}$$

Insert the values for the concentrations, punch a few keys on your calculator, and you should get

$$K_A^{298} = 5.0000 \times 10^{-4}/(3.2200 \times 10^{-3} \times 0.4995) = 3.1087 \times 10^{-1} \approx 0.311$$

This is the answer (A).

Problem 7.5 | The heat is on.

(B) If the temperature of the compartment is raised to 40.0°C the $\Delta G^{\ominus}{}_{40}$ for the ligand–receptor interaction is found to be +6.67 kJ. Use this data to calculate how much ligand is bound to the receptor.

7

» **Solution B – Strategy and Calculation**

Your "link" here is $\Delta G_{40}{}^{\ominus}$, the standard Gibbs energy for the ligand–receptor association at $T_2 = 40.0°C$. From the standard Gibbs energy you will calculate the association constant, K_A. Then, from the K_A, you will find the ligand–receptor concentration, $[R{:}L]$, at this temperature. Let us start with a good equation:

$$\Delta G_{40}^{\ominus} = -RT \ln K_A^{40}$$

We have to invert the equation to find K_A, and using the fact that exp[argument] is the inverse function to ln[argument] we write

$$K_A^{40} = \exp[-\Delta G_{40}^{\ominus}/RT]$$

Insert the numbers for $\Delta G^{\ominus}{}_{40}$, R, and T (convert it to degrees Kelvin!) and you should get

$$K_A^{40} = \exp[-6.67 \times 10^3 \,[\text{J}]/8.314\,\text{J K}^{-1} \times (273.2 + 40.0)\,[\text{K}]]$$

$$K_A^{40} = \exp[-2.56] = 7.72 \times 10^{-2} \approx 0.077$$

Done for the association constant at 40.0°C, $K_A{}^{40}$. Let us re-write the expression for the equilibrium constant:

$$K_{eq} = [R{:}L]/[R]_{free} \times [L]_{free} \tag{7-19}$$

The question here is: What is the equilibrium concentration of receptor–ligand complex, $[R{:}L]$, at $T_2 = 40.0°C$? We do not know $[R{:}L]$ or $[R]_{free}$ or $[L]_{free}$. We do know, however, $[R]_{tot}$ and $[L]_{tot}$ values – they are the same at each temperature.

What are we going to do now? The idea is to replace $[R{:}L]$ and $[R]$ in the expression for the association constant with something we already know or with something we can calculate. To do this we are going to use the substitution trick, *little big trick* #2, and we are going to use it at least twice. You should write the expression for the total concentration of ligand

$$[L]_{tot} = [L]_{free} + [R{:}L] \qquad (7\text{-}20)$$

and use it to express $[R{:}L]$:

$$[R{:}L] = [L]_{tot} - [L]_{free} \qquad (7\text{-}21)$$

Note that the unknown quantity $[R{:}L]$ is now expressed with the help of one known quantity, $[L]_{tot}$, and one quantity we are going to calculate, $[L]_{free}$. You may now re-write the expression for the total concentration of receptor in the solution:

$$[R]_{tot} = [R]_{free} + [R{:}L] \qquad (7\text{-}22)$$

And now try to express some of the unknown quantities with the help of the known quantities, for example, and the quantities we are going to calculate. In the expression above we defined $[R{:}L]$ with the help of $[L]_{tot}$ and $[L]_{free}$. Let us make this substitution in the expression for total concentration of the receptor:

$$[R]_{tot} = [R]_{free} + [L]_{tot} - [L]_{free} \qquad (7\text{-}23)$$

Now move the unknown quantity, $[R]_{free}$, to the left side:

$$[R]_{free} = [R]_{tot} - [L]_{tot} + [L]_{free} \qquad (7\text{-}24)$$

Let us pause to see what we have done. In (7-21) we have expressed $[R{:}L]$, an unknown quantity, with $[L]_{tot}$, which is known, and $[L]_{free}$, which we are going to calculate. In (7-24) we have expressed $[R]_{free}$, another unknown quantity, with $[R]_{tot}$ and $[L]_{tot}$, which are both known and, again, with $[L]_{free}$, which we are going to calculate.

So let us calculate the $[L]_{free}$ value. We should re-write the expression for the association constant

$$K_{eq} = [R{:}L]/[R]_{free} \times [L]_{free}$$

and make the two substitutions we have made in (7-21) and (7-23):

$$K_{eq} = ([L]_{tot} - [L]_{free})/([R]_{tot} - [L]_{tot} + [L]_{free}) \times [L]_{free}$$

When you multiply what has to be multiplied in the denominator and bring it up to the left side of the equation you will have

$$K_{eq} \times [R]_{tot} \times [L]_{free} - K_{eq} \times [L]_{tot} \times [L]_{free} + K_{eq} \times [L]_{free}^2 = [L]_{tot} - [L]_{free}$$

Clearly, this is a quadratic equation in $[L]_{free}$; $[L]_{free}$ is our x and we will make this substitution in the equation

$$K_{eq} \times x^2 + (K_{eq} \times [R]_{tot} - K_{eq} \times [L]_{tot} + 1) \times x - [L]_{tot} = 0$$

Recalling that the generalized form of the quadratic equation and its solution are given by the following expression

$$Ax^2 + Bx + C = 0 \quad \text{and} \quad x_{1,2} = \{-B \pm \sqrt{(B^2 - 4AC)}\}/2A$$

we get, after inserting the given values, that is, $K_{eq} = 0.077$, $[R]_{tot} = 3.72 \times 10^{-3}$, and $[L]_{tot} = 0.5000$,

$$-0.077x^2 + 0.9618x - 0.500 = 0 \quad \text{and}$$

$$x_{1,2} = \{-0.96179 \pm \sqrt{(0.92504 + 0.1540)}\}/0.1540 = (-0.96179 \pm 1.0387)/0.1540$$

$$x_1 = 0.49986$$

Do not round this number up; you will see in a moment why.

This is the concentration $[L]_{free}$ at 40°C. Note that the other solution of the quadratic equation, $x_2 = (-0.962 - 1.039)/0.154 = -12.99$ is a meaningless value so we ignore it. Now it is easy to find the concentration of the receptor–ligand complex at 40.0°C; it is the difference between the total ligand and free ligand:

$$[R{:}L] = [L]_{tot} - [L]_{free} = 0.5000 - 0.49986 \approx 1.400 \times 10^{-4} \text{ M}$$

An afterthought: Let us think a moment about these numbers: at 25°C $[R{:}L]$ is 5.0×10^{-4} M while at 40°C it is 1.40×10^{-4} M, or about 3.6 times lower. What can you conclude about the ligand–receptor interaction from these numbers? I would say that the temperature is *not good* for the R:L complex – but the precision of the data is not sufficient to fully support this statement.

Problem 7.6 | A disorderly chamber.

Assuming that the $\Delta H^{\ominus}$ for the ligand–receptor binding in this experimental setup does not change significantly between the room temperature and 40.0°C, calculate the $\Delta S^{\ominus}$ for the $R + L = R{:}L$ interaction.

» Solution C – Strategy and Calculation

You have solved this type of riddle before (Problem 7.1(B)) so we will not spend much time here. First, you express the standard Gibbs energies using enthalpy, the two temperatures, and entropy:

$$\Delta G_{T1}^{\ominus} = \Delta H_{T1}^{\ominus} + T_1 \Delta S^{\ominus} \quad \text{(A)}$$

$$\Delta G_{T2}^{\ominus} = \Delta H_{T2}^{\ominus} + T_2 \Delta S^{\ominus} \quad \text{(B)}$$

For $T_2 = 40.0°C = 40.0 + 273.2 = 313.2$ K, the $\Delta G^{\ominus}{}_{40}$ is given; it is 6.67 kJ. For room temperature, $T_1 = 25°C = 25.0 + 273.2 = 298.2$ K, you will calculate $\Delta G^{\ominus}{}_{25}$ from the value of $K_A{}^{25}$ value which you calculated in part (A):

$$\Delta G_{25}^{\ominus} = -RT \ln K_A^{25}$$

$$\Delta G^{\ominus} = -8.314 [\text{J K}^{-1}] \times 298.2 \, [\text{K}] \times \ln 0.311 = +2895.6 \, \text{J} = +2.90 \, \text{kJ}$$

Now insert these values in (A) and (B) and subtract (B) from (A):

$$2.90 \times 10^3 - 6.67 \times 10^3 = \Delta H_{T1}^{\ominus} - \Delta H_{T2}^{\ominus} - 298.2 \times \Delta S^{\ominus} - (-313.2 \times \Delta S^{\ominus})$$

Cancel the $\Delta H^{\ominus}{}_{T1}$ and $\Delta H^{\ominus}{}_{T2}$ as they are presumed to be equal and solve the equation for $\Delta S^{\ominus}$:

$$\Delta S^{\ominus} = (2.90 \times 10^3 - 6.67 \times 10^3) \, [\text{J}]/(298.2 - 313.2) \, [\text{K}]$$

$$\Delta S^{\ominus} = +251 \, \text{J K}^{-1}$$

Positive standard entropy of the ligand–receptor association in this experimental setup means that there is a lot more free ligand and receptor than the R:L complex.

A comment: This was quite a problem, wasn't it? If you understand most of the steps we went through you should feel confident about the questions, riddles – and real-life problems – involving concentrations, equilibrium constants, and standard thermodynamic functions, G, H, S of chemical reactions. And this is probably the major area of physical chemistry. I am proud of you – you have been a good companion on a long and difficult journey. Now I would like to see you becoming more of a guide than a tourist on this journey.

A note on the thermodynamics of genetic code: As you know the blueprint for your body – the whole of you – is stored in the deoxyribonucleic acid, DNA, molecules in your body. You may think of a DNA molecule as a string of beads. There are four kinds of beads in DNA and we label them as A, T, G, C. If you know how to read the beads, three at a time, you should be able to make and put together other molecules in your body. The string – the biologists prefer to call it a *strand* – is an important molecule, like a phone book or an address book. So to make sure that it does not get damaged or that the ordering of the beads does not get scrambled nature uses two same strands of beads in each DNA molecule, only the beads are ordered in the opposite sense: they go as A → A→ T → T → in one strand and as ← T ← T ← A ← A in the other. The biologists call this molecule the double-stranded DNA. A DNA is a long molecule – there are about 3.2 billion of these beads in each strand of a human DNA. The beads in the two strands hold together not by covalent chemical bonds but by weak interactions; we call them *hydrogen bonding*, HB, interactions. You may think of two DNA strands as of two parts of a Velcro® tape used to fasten a jacket or a shoe; they hold tightly together yet you can separate them. You can pull them apart – and put them back together in the same order.

The strands in a double-stranded DNA have to be separated in order to be read – and they are being read, by *reader molecules*, every moment of our lives.

So unzipping of the two strands and zipping them back together is an important and ongoing process. In nature, this is carried out by special molecules. In a laboratory we use mild heat, 50–100°C, to break the hydrogen bonds and separate the two strands. When the two strands are separated we can see individual beads by shining UV light on them. So the method of *UV spectrophotometry* is often used to observe separation of the DNA strands. We will talk more about this method later (Chapter 12).

Problem 7.7 | (Reading) Unzipping the book of life.

A sample of a primate mitochondrial DNA, "mtDNA." is placed into a temperature-controlled heating block and subjected to heating [3]. During this procedure the DNA undergoes a reversible transformation ("reaction") from a double-stranded (ds) to a single-stranded (ss) molecule:

$$dsDNA \leftrightarrows ssDNA$$

At each temperature we take a smaller sample and measure its UV absorbance at 260 nm; the more UV light is absorbed the more DNA is separated into single strands. The absorbance at 37.0°C tells us that 92.7% of the total DNA is still double stranded.

(A)   Find the equilibrium constant for the ds $\leftrightarrows$ ss change and the change of the standard Gibbs energy for this process, $\Delta G^{\ominus}_{37}$.

(B)   When the temperature increased to 61.8°C, however, the double-stranded heteroduplex mtDNA (heteroduplex = made of two different strands) undergoes extensive dissociation ("melting"); the UV absorbance shows that 79.7% of the mtDNA becomes single stranded. Find the equilibrium constant and the standard Gibbs energy at that temperature, $\Delta G^{\ominus}_{62}$.

(C)   Using the data from (A) and (B) and assuming that the standard enthalpy for this process remains about the same over the 37.0–61.8°C temperature range, estimate the standard entropy, $\Delta S^{\ominus}$, for the process of *unzipping* the DNA.

» **Solution – Strategy A and B**
» **Assumption #1**

The mitochondrial DNA we are talking about here can be double stranded or separated into single strands. At some lower temperature, typically, +4°C, all of the DNA is double stranded. How much of DNA? You may use *little big trick #3* and write *one* mole. So 1 mol, or 100%, of the DNA is dsDNA. Since there are no other forms of DNA, besides the double-stranded and single-stranded ones, present in this sample, this means that the percentage of ssDNA is zero, 0%. At some higher temperature, say +95°C, it is the other way around: all of the DNA is unzipped and you have 0% of dsDNA and 100% of ssDNA. At all other temperatures between +4 and +95°C the DNA in your sample is a mixture of *ds* and *ss* forms.

» Assumption #2

At each temperature the DNA sample is kept long enough to reach an equilibrium between the *ds* and *ss* forms. This allows us to calculate the equilibrium constant and all thermodynamic functions at that temperature.

I think we have all the knowledge that we need so – let us calculate some numbers.

» Solution – Calculation a and b

The equilibrium constant for the *ds* ⇌ *ss* transformation is the ratio of the amount (or concentration) of the *ss* form divided by the amount of the *ds* form:

$$K_{eq} = [product]/[reactant] = [ssDNA]/[dsDNA]$$

Note: If you are thinking that 1 mol dsDNA will give 2 mol ssDNA, when completely unzipped, you are completely right. The expression for the equilibrium constant would then read $K_{eq} = [ssDNA]^2/[dsDNA]$. However, in this type of problem we are not concerned with the actual number of molecules – or moles – but with the relative amount, the percentage, of the DNA in *ds* form and the percentage of the DNA in *ss* form. (This is a little bit like when you have a bag with 2 blue candies and 3 yellow candies and all you need to know is their ratio, blue/yellow = 2/3, not how many candies there actually are in the bag.)

The UV absorbance tells us there is 92.7% of the DNA in *ds* form so the rest – what is left of 100% – must be ssDNA; we write 100.0% – 92.7% = 7.3% ssDNA. Insert these numbers into the equation for equilibrium between the total amounts of different forms of DNA and you get

$$K_{37} = 7.3/92.7 = 0.079$$

The standard Gibbs energy for this process is given as

$$\Delta G^{\ominus} = -RT \ln K_{eq}$$

$$\Delta G^{\ominus}(37.0°C) = -8.31 \, [J \, K^{-1} \, mol^{-1}] \times (273.2 + 37.0) \, [K] \times \ln 0.079$$

$$\Delta G^{\ominus}(37.0°C) = -8.31 \, [J \, K^{-1} \, mol^{-1}] \times 310.2 \, [K] \times (-2.54)$$

$$\Delta G^{\ominus}(37.0°C) = +6.55 \times 10^3 \, J \, mol^{-1} = 6.6 \, kJ \, mol^{-1}$$

Now the higher temperature. The UV absorbance for the DNA sample at 61.8°C tells you that most of the DNA, 79.7%, has been converted into single-stranded form. How much of the dsDNA is left there? The %(ds) = 100.0% − 79.7% = 20.3%. So now numerator is the larger number:

$$K_{62} = 79.7/20.3 = 3.93$$

$$\Delta G^{\ominus}(62°C) = 8.31\,[\mathrm{J\,K^{-1}\,mol^{-1}}] \times (273.2 + 61.8)\,[\mathrm{K}] \times \ln 3.93$$

$$\Delta G^{\ominus}(62°C) = -3809\,\mathrm{J\,mol^{-1}} = -3.8\,\mathrm{kJ\,mol^{-1}}$$

Comment: Take another look at these two standard Gibbs energies: +6.6 kJ at 37.0°C (the human body temperature) and −3.8 kJ at 61.8°C (the temperature of a hot tea). The positive $\Delta G^{\ominus}$ at 37.0°C tells that the process is *not spontaneous*; it has to be driven. At higher temperature it is all downhill, the $\Delta G^{\ominus}$ is negative, and the rest of dsDNA will unzip spontaneously whether you want it to happen or not.

» Solution – Strategy and Calculation C

Let us read Part (C) again:

(C) Using the data from (A) and (B) and assuming that the standard enthalpy for this process remains about the same over the 37.0–61.8°C temperature range, calculate the standard entropy, $\Delta S^{\ominus}$, for the process of *unzipping* the DNA.

You may recall that we had this type of problem before ("What the old malt is actually made of"); it is based on the assumptions that $\Delta H^{\ominus}$ and $\Delta S^{\ominus}$ do not change significantly over the given temperature range. You may remember this or you may want to spell it out:

$$\Delta H^{\ominus}_{37} \approx \Delta H^{\ominus}_{62} = \Delta H^{\ominus} \quad \text{and} \quad \Delta S^{\ominus}_{37} \approx \Delta S^{\ominus}_{62} = \Delta S^{\ominus}$$

Note that these are the *standard enthalpy* and *standard entropy;* the actual enthalpy and entropy can change quite a bit between different temperature points. Using these two assumptions you proceed in the following way:

- You write two equations of the type $\Delta G^{\ominus} = \Delta H^{\ominus} - T\,\Delta S^{\ominus}$ for two given temperatures (37.0 and 61.8°C)
- Subtract one equation from the other and cancel the $\Delta H^{\ominus}$
- Insert the values for two $\Delta G^{\ominus}$ and two temperatures and
- Solve the remaining equation for $\Delta S^{\ominus}$

The relation between Gibbs energy, enthalpy, and entropy is the most important thermodynamic equation:

$$\Delta G^{\ominus} = \Delta H^{\ominus} - T\Delta S^{\ominus}$$

So for the two temperatures, $T_1 = 37.0°C + 273.2\,\mathrm{K} = 310.2\,\mathrm{K}$ and $T_2 = 61.8°C + 273.2\,\mathrm{K} = 335.0\,\mathrm{K}$, which you are going to convert to degrees Kelvin, you will have

$$\Delta G^{\ominus}(37.0°C) = \Delta H^{\ominus} + 310.2\,[K] \times \Delta S^{\ominus} \qquad (A)$$

$$\Delta G^{\ominus}(61.8°C) = \Delta H^{\ominus} + 335.0\,[K] \times \Delta S^{\ominus} \qquad (B)$$

This is yet another application of *little big trick* #5: Subtract (B) from (A) and cancel the same terms:

$$\Delta G^{\ominus}(37.0°C) - \Delta G^{\ominus}(61.8°C) = (\Delta H^{\ominus} + 310.2 \times \Delta S^{\ominus}) - (\Delta H^{\ominus} - 335.0 \times \Delta S^{\ominus})$$

(We do not really need parentheses but they make the expression above easier to read.) Clearly $\Delta H^{\ominus}$ can be eliminated: do it. Then insert the values (and units!) for the $\Delta G^{\ominus}$ we have calculated above and solve the equation for $\Delta S^{\ominus}$:

$$\Delta S^{\ominus} = \{+6.6\,kJ\,mol^{-1} - (-3.8)\,kJ\,mol^{-1}\}/\{310.2\,[K] - 335.0\,[K]\}$$

Caution: If you keep in mind that the $\Delta G^{\ominus}$ are given in kJ, that is, in 1000 joule units, you will get

$$\Delta S^{\ominus} = -419\,J\,K^{-1}\,mol^{-1}.$$

Comment: This is a large negative entropy and to achieve this transformation a larger *positive* entropy will have be created somewhere else. (This is for example when you are putting things *in order* – books, clothes, etc. – you are reducing the disorder in the space around you. But to do this you have to spend a lot of energy and time and create a large positive surplus of entropy.) Apparently nature does not *like* when the two strands in your DNA get separated.

7.2 Acids and Bases

One of the many different chemical reactions occurring in nature or in the laboratory is arguably the most important. It takes place between many different pairs of compounds yet it always follows the same general scheme:

$$HA \rightarrow A^- + H^+ \qquad\qquad\qquad\qquad (7\text{-}25)$$

Here A can be any atom or molecule and I will say more about it in a moment. H^+ is a hydrogen atom stripped of its electron; we call it a *proton*. So if "something" – we label it HA in the scheme above – can give a proton away we call it an *acid*. HA is an acid. This makes A^- a *base*, or, since it is related to HA through exchange of a proton, we often call it the *conjugate base*. (Note: *Conjugate* – not *conjugated*, which has an entirely different meaning in chemistry.)

So what is an acid? Water, sulfuric acid (*danger!*), vinegar (acetic acid), or deoxyribonucleic acid, DNA, the molecule containing your genetic code – they are all acids. On the other hand, boron trifluoride, BF_3, carbon monoxide, CO, or sulfur dioxide, SO_2, are not acids; none of these molecule can give a proton away. The acids that can give a proton away are also called Brønsted acids, in honor to Danish physical chemist Johannes Brønsted. (There are also Lewis acids.)

Like all other reactions, processes or interactions in equilibrium, an acid, a base, and a proton are connected through equilibrium constant; for water this will read

$$HOH = HO^- + H^+$$

$$K_a = [HO^-][H^+]/[HOH] \tag{7-26}$$

The acid–base equilibrium constants can be large, K_a ($CF_3SO_2OH/CF_3SO_3^-$) = 10^{+13} or very small numbers, K_a (CH_4/CH_3^-) = 10^{-55}, so it is customary to convert them to logarithms. Historically, a negative logarithm base 10, labeled p, has been used in acid–base chemistry. So we write $-lg_{10}K_a$ or just $-lg\ K_a = pK_a$. Similarly, $-lg_{10}[H^+] = pH$. (Do not confuse this p with the p meaning *pico*, that is, one trillionth part of something, e.g., pm = 10^{-12} m.)

The logarithmic form of the acid–base equilibrium constant for water will then be

$$pK_a = -lg[HO^-] - lg[H^+] - (-lg[HOH]) \tag{7-27}$$

It is customary to express $-lg[H^+]$ as pH which gives us the so-called Henderson–Hasselbalch equation, or "HH" equation, named after the American scientist Lawrence Henderson and Danish chemist Karl Hasselbalch:

$$pK_a = pH - lg[HO^-]/[HOH] \tag{7-28}$$

Note that pK_a is often an experimentally known value, pH can usually be measured and $lg([HO^-]/[HOH])$ is a *ratio* of the concentrations (activities, actually) of the base and the acid in this reaction, $[A^-]/[HA]$. Oftentimes we need just the base/acid ratio, rather than individual $[A^-]$ and $[HA]$ values which simplifies this type of calculations.

Acids and bases are treated in every chemical course so we will keep this section short; we will look at two examples only.

Problem 7.8 | You'll drink up vinegar? (W. Shakespeare, Hamlet, Act V, Scene 1).

Acetic acid, $pKa = 4.73$, is added to aqueous solution at pH 3.9. What percentage of acetic acid, CH_3COOH, has dissociated in this solution to give CH_3COO^- and H^+?

》 **Solution – Strategy and Calculation**

The HH equation is a golden tool in acid–base reactions and calculations – use it! You may want to first write the dissociation reaction for acetic acid:

$$CH_3COOH = CH_3COO^- + H^+$$

The equilibrium constant is the ratio of concentrations of products (everything on the right side of the equation) and the reactants (everything on the left side of the equation):

$$K_a = [CH_3COO^-][H^+]/[CH_3COOH]$$

And if you apply the −log function throughout the expression for equilibrium you will get the HH equation for acetic acid:

$$pK_a = pH - lg[CH_3COO^-]/[CH_3COOH]$$

You are given pK_a and pH values; the question is how much has the acid dissociated in this solution. In other words the question is, How much of CH_3COOH got converted to CH_3COO^-? This may again look like one of "those" riddles where too many questions are asked and too little information given. You do not know either the initial or the equilibrium concentrations of CH_3COOH; neither do you know the equilibrium concentration of CH_3COO^-.

What do you do when no amount of the substance – kilograms, pounds, moles – is given? You use *little big trick #3*; assume that you started with *one mole*.

So here is 1 mol of CH_3COOH. When you place it in water the acid dissociates to give CH_3COO^- and H^+ in equal number of moles. How much of CH_3COO^- is there at equilibrium? You do not know; nobody does. So you write x; x moles of CH_3COO^- (and x moles of H^+). This leaves us CH_3COOH diminished by x moles and you should write $(1-x)$ moles of CH_3COOH left in equilibrium. Now you have everything you need to calculate x, the part of initial CH_3COOH that got converted into CH_3COO^- and H^+. Make the following substitutions in the HH equation: x for $[CH_3COO^-]$ and $[1-x]$ for CH_3COOH. Now the equation will read

$$pK_a = pH - lg\{x/(1-x)\}$$

Let us insert the numbers we were given and see what we can do next:

$$4.73 = 3.9 - lg\{x/(1-x)\}$$

$$-0.83 = lg\{x/(1-x)\}$$

You need to calculate the argument of the log function so you will use the function inverse to lg, which is the power to ten:

$$10^{-0.83} = 10^{lg\{x/(1-x)\}}$$

The two inverse functions, power to ten and logarithm base 10, cancel each other, that is, $10^{lg\{something\}} = something$, so you will get

$$10^{-0.83} = x/(1-x)$$

which further gives

$$0.148 \times (1-x) = x$$

And you solve it for x:

$$x = 0.148/(1-0.148) = 0.174$$

And – voilà – you have it: from the initial 1 mol of CH_3COOH only 0.174 moles or 17.4% dissociated to CH_3COO^- and H^+.

A comment on acid dissociation: I was thinking of the number we just calculated: 17.4%. Seventeen point four percent of the initial amount of acetic acid. How much was that? One mole, hundred moles, or 0.0001 mole – it does not matter; all we need is the *ratio* of dissociated form, CH_3COO^-, to undissociated form, CH_3COOH. The amount of 17.4% is not a big number, it is less than 1/5 of what we had at the beginning. This is telling me that CH_3COOH is not a strong acid; it does not give H^+ away easily. (We humans have known this for a long time: this is why we put vinegar and not sulfuric acid into salads and other dishes.) I should also mention that all this is happening in an aqueous solution at pH 3. This is 4 logarithmic units, or $10^4 = 10,000$, more acidic than pure water. I just did a quick calculation: if you put acetic acid into pure water at pH 7 then over 99% of the initial CH_3COOH will dissociate into CH_3COO^- and H^+.

Let us look at the pK_a of acetic acid: $pK_a \approx 4.7$ (Fig. 7.2). When pH of the solution is 7, that is, higher than pKa, most of the acid is in the dissociated form, CH_3COO^-. On the other hand when pH of the solution is lower than pK_a, like in this problem, pH = 3.9, most of the CH_3COOH remains undissociated. It seems to me there is a lesson to be learned here: the higher the pH of the medium the more dissociated (anionic, basic, deprotonated) the acid–base molecule. At lower, more acidic pH, the acid remains mostly undissociated, as acid. The "sweet spot" is when pKa equals pH: CH_3COO^- and CH_3COOH are then *equal*, that is, 50:50. This lesson applies nicely to the acid–base reactions of biological molecules as the next riddle tells us.

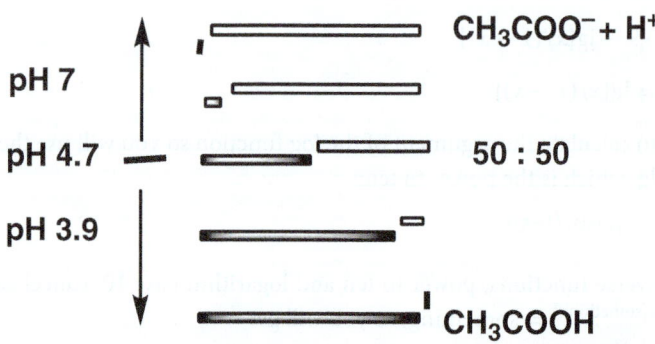

Fig. 7.2 The ratio of undissociated form, CH_3COOH, to anion, CH_3COO^-, exponentially decreases as pH of the solution gets higher

| Problem 7.9 | Thanksgiving [4, 5], Turkey, Tryptophan. |

After a hefty bite into your grandma's Thanksgiving turkey, the meat (hopefully chewed up) proceeds to your stomach. There, under the action of gastric juices, pH 1.8, it is degraded into peptides and amino acids. Next, the partially processed food enters the so-called duodenum tract, where pH jumps by full four logarithmic units to pH 6.2. Given that a free tryptophan has three acid–base centers, what is its predominant form in stomach and what in duodenum?

» Solution – Strategy and Calculation

This is a big riddle involving a complex molecule so I suggest we pull out some data and draw few formulas. Tryptophan, Fig. 7.3, is an *essential* amino acid (our bodily biochemistry is unable to make it from scratch so we have to ingest it) found in larger quantities in turkey meat. It also tends to make us a little drowsy and a little slow. "Because it is found in turkeys" goes one "scientific explanation." Like all amino acids tryptophan contains a carboxylic group, $-COOH$, an amino group, $-NH_2$, and, in the heterocyclic indole ring, a pyrrolo NH group.

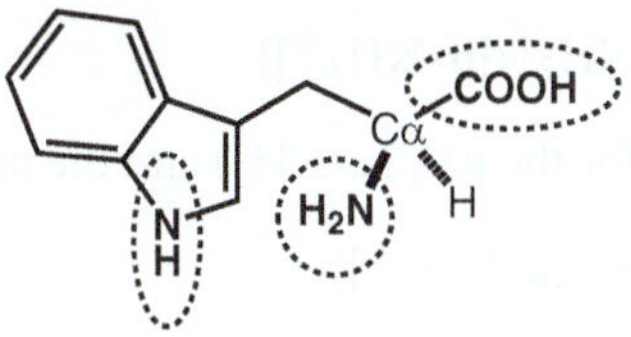

Fig 7.3 Tryptophan: mid-range acid–base groups

Each group can give away or accept protons (the α in front denotes that both groups are on the alpha-carbon atom, $C\alpha$):

$$\alpha\text{-COOH} = \alpha\text{-COO}^- + H^+ \quad pK_a = 2.38$$

The protonated α-amino group gives a positively charged acid: $\alpha\text{-NH}_3^+ = \alpha\text{-NH}_2 + H^+$ $pK_a = 9.39$.

There are also the acid–base dissociation reaction of protonated indole amino group and the acid–base dissociation of the α-amino group; however, both are way out of the pH range we are considering now: pK_a (indole-NH_2^+/indole-NH) –2 and pK_a (R-α-NH_2/R-α-NH^-) +35. This leaves you with the α-carboxylic acid group and α-amino group acid–base equilibria. If you think of the model for ligand–receptor reactions then you may consider H^+ as a ligand and tryptophan as a receptor with two ligand binding sites, α-COO and α-NH_2 groups. Let us now calculate some numbers.

You follow the turkey's journey through your body, with the first stop in the stomach, where pH is 1.8. In the strongly acidic gastric juices the α-amino group will accept a proton, so we will write an *association* reaction:

$$\alpha\text{-NH}_2 + \text{H}^+ = \alpha\text{-NH}_3{}^+$$

The equilibrium constant for this reaction is

$$K = [\alpha\text{-NH}_3{}^+]/[\alpha\text{-NH}_2][\text{H}^+]$$

If you apply a negative logarithm throughout you will get

$$-\lg K = -\lg\{[\alpha\text{-NH}_3{}^+]/[\alpha\text{-NH}_2][\text{H}^+]\}$$

The pK_a number we are given, $-\lg K_a = pK_a = 9.39$, is for an acid–base reaction which is a *dissociation* reaction. Like in the case of cortisol–glucocorticoid ligand–receptor reaction, Problem 7.2, the relation between the two equilibrium constants is $K = 1/K_a$ (keep in mind that the suffix a in K_a means *acidic*, not association). So you should write

$$-\lg(1/K_a) = -\lg\{[\alpha\text{-NH}_3{}^+]/[\alpha\text{-NH}_2][\text{H}^+]\}$$

You may use the following property of logarithms $\lg(1/x) = -\lg x$ and re-write the previous equation as

$$-\lg K_a = \text{pH} - \lg\{[\alpha\text{-NH}_2]/[\alpha\text{-NH}_3{}^+]\}$$

Insert the numbers given for the pK_a ($\alpha\text{-NH}_2$) and the pH in stomach and you get

$$9.39 = 1.8 - \lg\{[\alpha\text{-NH}_2]/[\alpha\text{-NH}_3{}^+]\}$$

From here you get for the base/acid ratio:

$$[\alpha\text{-NH}_2]/[\alpha\text{-NH}_3{}^+] = 10^{-7.59} = 2.57 \times 10^{-8}$$

This is a very small number. And what is it telling us? It is telling us that, in stomach, where pH is 2, the ratio of protonated α-amino group, $\alpha\text{-NH}_3{}^+$, to the unprotonated form, $\alpha\text{-NH}_2$, is less than 3 parts in a hundred million, or nearly nothing. The α-amino group in a free tryptophan in gastric juices is pretty much completely protonated.

And what about the α-COOH group? This is even simpler: you will write the dissociation reaction and use the HH equation to find the $\alpha\text{-COO}^-/\alpha\text{-COOH}$ ratio:

$$\alpha\text{-COOH} = \alpha\text{-COO}^- + \text{H}^+$$

$$K_a = [\alpha\text{-COO}^-][\text{H}^+]/[\alpha\text{-COOH}]$$

and the HH form of the equilibrium equation reads:

$$pK_a = \text{pH} - \lg\{[\alpha\text{-COO}^-]/[\alpha\text{-COOH}]\}$$

Now insert the numbers for pK_a ($\alpha\text{-COOH}/\alpha\text{-COO}^-$) = 2.38 and pH = 1.8, raise the equation through to the power of ten, and you get

$$[\alpha\text{-COO}^-]/[\alpha\text{-COOH}] = 10^{-(2.38-1.8)} = 10^{-0.58} = 0.26$$

What is this number telling us? It is telling us that in gastric juices, in only about 1/4 (0.26 = 26%) of tryptophan molecules the α-carboxylic group is dissociated and in 3/4 molecules it remains in protonated form. The predominant form of tryptophan at pH 1.8 is therefore a charged species, a *cation*; Fig. 7.4.

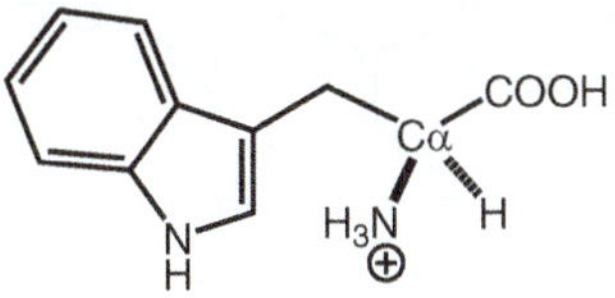

Fig. 7.4 Tryptophan cation in the gastric pH domain

Now the next stop on the journey: the duodenum, the first section of the small intestines. The acid–base conditions there are drastically different than in the adjoining stomach, pH = 6.2. Let us put these two numbers, pH 1.8 and pH 6.2, into perspective. I will subtract the second from the first and express the result as a negative power of ten. Why? Because what you are doing now is the *opposite* procedure of calculating pH from a given activity (concentration) of H^+ ions:

$$\Delta pH = 1.8 - 6.2 = -4.4$$

$$10^{-\Delta pH} = 10^{4.4} = 2.512 \times 10^{+4} = 25,120$$

The gastric juices in stomach are 25,000 more acidic than the "next door" duodenum. It is truly amazing how our bodies do these tricks.

From now on the calculations are going to be very easy – you use the formulas derived above and only insert the value for the pH in duodenum. For the $[\alpha\text{-}NH_3^+]/[\alpha\text{-}NH_2]$ acid–base pair the HH equation will read

$$9.39 = 6.2 - \lg\{[\alpha\text{-}NH_2]/[\alpha\text{-}NH_3^+]\}$$

From here you get for the base/acid ratio

$$[\alpha\text{-}NH_2]/[\alpha\text{-}NH_3^+] = 10^{-3.19} = 6.45 \times 10^{-4}$$

This is about six parts per ten thousand, a much bigger number than before but still a fairly small number. Even in duodenum, at pH = 6.2, most of the α-amino group is in protonated form, $\alpha\text{-}NH_3^+$. And now let us see what happens to α-carboxyl group in duodenum:

$$[\alpha\text{-}COO^-]/[\alpha\text{-}COOH] = 10^{-(2.38-6.2)} = 10^{+3.82} = 6607$$

This is radically different than in stomach; in duodenum the concentration of tryptophan with dissociated α-carboxylic group is almost seven thousand times larger than the concentration of the undissociated form. The predominant form of the

carboxylic group is therefore *anion*. The free tryptophan in this form has one positively charged group, α-NH$_3^+$, and one negatively charged group, α-COO$^-$, so it is
a double-charged ion, a *zwitterion*, Fig. 7.5.

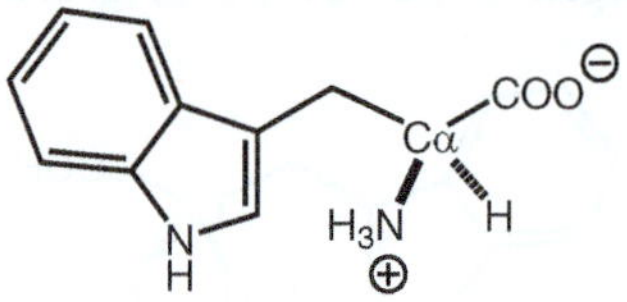

Fig. 7.5 Tryptophan zwitterion

7

References

Cortisol Receptor
1. Eliard PH, Rousseau GG (1984) Thermodynamics of steroid binding to the human glucocorticoid receptor. Biochem J 218:395–404.
2. Oelkers W (1997) Adrenal Insufficiency. N Engl J Med 335:1206–1212.

ΔH and ΔS of DNA Melting
3. Brown WM, George M Jr, Wilson A (1979) Rapid evolution of animal mitochondrial DNA.
   Proc Natl Acad Sci USA 76:1967–1971

Thanksgiving Day
4. http://en.wikipedia.org/wiki/Thanksgiving/, Accessed July 31, 2009
5. http://www.si.edu/Encyclopedia_SI/nmah/thanks.htm/, Accessed July 31, 2009

Table III

Summary of mixtures and chemical thermodynamics – what have we learned in this section?

Review the material we have covered, write down the new words and describe them. Then re-write these descriptions in symbols and numbers.

Words and Phrases	Symbols, Formulas and Numbers
Percent concentration	#g solute in 100 mL solution, % conc.
Molar concentration	# moles solute in 1,000 mL solution; M, mol L^{-1}
Molal concentration	# moles solute in 1.0 kg solvent, m
Molar fraction	$x_1 = n_1/n_{tot}$; $0.0 = x_1 = 1.0$; $n =$ # moles
Partial molar volume	Molar volume for specific solvent mixture
Chemical reaction	$\alpha A + \beta B \rightarrow \gamma C + \delta D$
Reaction quotient	$Q = a(C)^\gamma\ a(D)^\delta /a(A)^\alpha\ a(B)^\beta \approx [C]^\gamma [D]^\delta /[A]^\alpha [B]^\beta$
Activities and concentrations	$a = \gamma\ c_M \approx c_M$ or [M], for $c < 0.01$
Chemical equilibrium, equilibrium constant	$K_{eq} \approx [C]^\gamma [D]^\delta /[A]^\alpha [B]^\beta$
Standard Gibbs free energy	$\Delta G^\ominus = - RT \ln K_{eq}$
Receptor–ligand interaction; Equilibrium constant	$R + L \rightarrow R{:}L$ $K_{eq} = [R{:}L]/[R]\,[L]$
Receptor–many-ligand interaction; Equilibrium constants	$R + L \rightarrow R{:}L + L \rightarrow R{:}L_2 + L \rightarrow R{:}L_3$ $K_A(3)K_A(2)K_A(1) = [R{:}L_3]/[R][L]^3$
Diffusion-restricted equilibria	$K_{eq} = [R{:}L]/[R]_{free}\ [L]_{free}$
Acid–base reaction	$HA \rightarrow A^- + H^+$
pH of a solution (approximate)	$pH = -lg_{10}\ [H^+]$; where $[H^+] =$ molar conc. of H^+
Concentration of H^+	$[H^+] = 10^{-pH}$
Acid–base equilibrium (approximate)	$K_a = [H_3O^+]\ [A^-]/[HA][H_2O] \approx [H^+][A^-]/[HA]$
Henderson–Hasselbalch equation	$pK_a = -lg_{10}\ K_a = pH - lg\ \{[A^-]/[HA]\}$, where $lg = lg_{10}$

Part IV

Ionic Properties and Electrochemistry

8   Ions 99

9   Electrochemistry 111

8 Ions

According to Greek mythology the daughter of Erechtheus, the king of Athens and the symbolic father of all Athenians, illegitimately conceived a child with god Apollo. She abandoned her son after birth expecting him to die. An orphaned, homeless person in low esteem, he found, much later and after many complications, his mother and reunited with her and subsequently achieved a great success and happiness in life. His name was ION.

From the story of Ion as told by Euripides.

In the previous two problems we stated that an acid dissociates to give a conjugate base, an *anion*, and a proton, a *cation*. Anions and cations are *ions*. When you dissolve a spoonful of sugar in a glass of water you get a solution of sugar molecules; they are same, in shape and size, as in a lump of sugar. However, a spoonful of salt, NaCl, dissolved in water will give you one spoonful of sodium ions, Na^+, and one spoonful of chloride ions, Cl^-. So you end up having two times as many "things." There are many other interesting properties ions have. Ions are present everywhere in nature and in our bodies, in our sweat, our blood, and our tears.

When working with ions you have to keep track of *three* of their properties:

(1) Ions do not come alone, that is, you may not have a jar with Na^+ ions and another jar with Cl^- ions. Ions are created when you dissolve a material that we call *ionic* or *ionizable*. Kitchen salt, NaCl – and most other salts, e.g., $MgSO_4$ or Epsom salt – will give ions; so will hydrochloric acid and many other compounds.

(2) Ions are charged particles (by *particles* I mean molecules, ions, electrons – pretty much anything very small) and as such they attract each other. How strongly they will attract each other depends on the kind of solvent they are in. Benzene, hexane, and other solvents made of C and H atoms only do not keep the ions well apart and cations (positively charged ions) and anions (negatively charged ions) unite and stay together like they do in a crystalline, solid state. If you put a few grains of salt, NaCl, into a beaker filled with hexane, C_6H_{12}, the salt will not dissolve no matter how hard you stir or warm up the solvent. We call

P-P. Ilich, *Selected Problems in Physical Chemistry*,
DOI 10.1007/978-3-642-04327-7_8, © Springer-Verlag Berlin Heidelberg 2010

hexane, benzene, and a number of solvents similar to hexane *nonpolar solvents*. Water is a good solvent for ions as it keeps each ion surrounded by solvent molecules (solvent dressed, Fig. 6.2(C)) and separated from each other. We call water, ethanol, acetic acid, formamide, and other solvents similar to water *polar solvents*. Many polar solvents can either give away or accept a proton so they act as weak to medium-strong acids and bases. This kind of knowledge can be very useful in a laboratory.

(3) Another property of ions is that when an ionic molecule is well dissolved, it falls apart into two, three, or more ions. Kitchen salt, as we said, will give two particles: Na^+ and Cl^-. Another salt, aluminum sulfate, $Al_2(SO_4)_3$, will give two Al^{+3} ions and three SO_4^{-2} ions: five particles in total. Now you have to be careful about this. When you dissolve 1 g of NaCl you will not get a total of 2 g of Na^+ and Cl^- ions; the total mass will still be no more than 1 g. You should think in terms of the number of particles. When you dissolve 1 mol of NaCl, which contains an Avogadro's number of NaCl molecules, you will get 1 mol of Na^+ ions and 1 mol of Cl^- ions: two moles in total. So it is the *number of moles* or the number of molecules, rather than the mass or volume, that will multiply in a solution of ions.

It is this last property, the number of particles in a solution, we will consider in the following example.

When you dissolve something in water, for example, salt, the solution has slightly different properties than pure water. For example, a salty solution boils at a higher temperature and freezes at a lower temperature than pure water. Physical chemists have established that these changes, from pure solvent to solution, depend on the number of particles of the solute and not so much on the type of solute. The common name for these effects is *colligative* properties of mixtures and solutions. Outside of a physical chemical laboratory those of us who live in colder climates are well familiar with the phenomenon of *freezing point depression* or *cryoscopy*. This is also a physical chemical experiment actively practiced by the fish swimming in arctic seas.

| Problem 8.1 | An arctic night in Coral Gables, FL. |

Following a whole day of heavy rain a spell of unusually cold air descends on Coral Gables, Florida (average yearly temperature 23.3°C), lowering the temperature to –1.7°C and turning the city streets into a skating rink. The city's Engineering Department summons for an emergency evening meeting. The plan is to cover the streets with some "ice-melting agent" and prevent the likely traffic carnage the next morning. The question is, What to use to melt the ice on the streets? A local construction material supplier offers salt, NaCl, at $54 per ton while the agrochemical factory just outside the town has a large supply of urea, $(NH_2)_2C=O$, at \$42 per ton. Fully aware of the urgency of the situation the members of the department are also trying to save the city's money and, keeping in mind the 30 square kilometers of the city's roads, they promptly decide to purchase 100 tons of the less expensive material. Barring the environmental impact difference between salt and urea, do you think this was a good decision? Support your argument by numbers.

» Solution – Strategy and Calculation

So here you are: a young employee in the city of Coral Gables Engineering Department, faced with an important decision. Guess what is going to help you? Your knowledge of physical chemistry.

This is a situation where you need to employ the colligative properties of water solution, specifically, to lower the freezing point of the water covering the streets. You check a physical chemistry textbook (or an Internet source) and find the following formula for the freezing point depression, ΔT_f:

$$\Delta T_f = K_f \times (\text{\# moles of solute}/1000\,\text{g solvent}) \tag{8-1}$$

Look carefully at this formula for all your answers are there. It tells you that the freezing point depression, ΔT_f, depends on K_f and the molality of the *defrosting agent* you are adding to the solution. K_f is an experimental constant, $1.86°C$ for water. It means that if you make one molal solution of *something* in water that water solution will freeze at $-1.86°C$ instead of at $0°C$. If you make a 5 molal solution, the water will freeze at $t = 5.0 \times (-1.86°C) = -9.3°C$; quite cold. So the more moles you add and dissolve in water the lower will be the freezing temperature of the solution. You cannot control the amount of solvent, that is, the frozen rain on the city streets (though a rough estimate can be made using the total street area and an average ice layer thickness) but you can easily calculate the *number of moles* you get for your money. Given that 1 ton $= 1,000$ kg and the molecular mass of NaCl is $58.44\,\text{g cm}^{-3}$ or $0.05844\,\text{kg m}^{-3}$, you get for number of moles of NaCl in 100 tons of salt

$$n(\text{NaCl}) = 100 \times 1,000.00\,[\text{kg}]/0.05844\,[\text{kg mol}^{-1}] = 1.711 \times 10^6\,\text{mol}$$

Hundred tons of urea, with the molecular mass of $0.06006\,\text{kg mol}^{-1}$, contain the following number of moles:

$$n(\text{CO(NH}_2)_2) = 100 \times 1,000.00\,[\text{kg}]/0.06006\,[\text{kg mol}^{-1}] = 1.665 \times 10^6\,\text{moles}$$

So from a 100 ton of either material you get few more moles of NaCl than of urea, or, to be precise, we write

$$n(\text{NaCl})/n(\text{CO(NH}_2)_2) = 1.711 \times 10^6 /1.665 \times 10^6 = 1.0277\text{ or }102.77\%$$

Subtract a 100 from it and you get a surplus of 2.77% for NaCl. But when you compare the prices, this looks very different:

price (NaCl)/price (urea) = \$54/\$42 = 1.2857 or 128.57%

Subtract a 100 from it and you get 28.57; a 28.57% higher price for salt. True, salt will give you few more moles – and therefore can be used to defrost a larger area, about 2.8% larger – but it is also almost 30% more expensive. It seems it is better to buy urea. Yes – but! Here a deeper knowledge and understanding of physical chemistry comes into play: NaCl is an ionic compound and urea is not. This means that each molecule of NaCl, when dissolved in water – that is, when it comes in contact with ice – will give one Na^+ particle and one Cl^- particle. When translated to 100 tons of salt this means you will get 1.711×10^6 mol $\times$ 2 = 3.42×10^6 mol of solute. This is a *lot* more than what you get from 100 tons of urea, as the following comparison shows:

$$n(Na^+, Cl^-)/n(urea) = 3.42 \times 10^6/1.665 \times 10^6 = 2.055 \text{ or } 205.5\%$$

When buying salt, at a 28.57% higher price, you get more than 205% the effect. So salt is less than a third more expensive but also more than twice as effective. Clearly – it is a much better choice as a defrosting agent. What now? Just as the meeting of the city's engineering department is about to be adjourned you walk back and say – "May I get your attention for a moment?"

A note on colligative properties: Salty water will not only freeze at a lower temperature than pure water, it will boil at a higher temperature too. Osmotic pressure is yet another effect observed for solutions, particularly concentrated solutions, when compared to pure solvents. The latter is a misnomer as no pressure, that is, force acting upon area, is involved. The effect of osmotic pressure is that very concentrated solutions in contact with pure solvent (or less concentrated solutions) get diluted by causing a net transfer of solvent; the process is driven by the difference in the Gibbs free energies between the two solutions. Though widely used in industry and laboratory – particularly osmotic pressure and freezing point depression – colligative properties are pure phenomena, observed and known for centuries if not longer, yet still without a comprehensive explanation involving the properties of matter and the nature of interaction between different phases.

8.1 Ion Activities

Ions, being electrically charged particles, conduct electricity. Accordingly, solutions of ions are called *electrolytes*. A solution of NaCl in water is an electrolyte; a solution of urea in water is a non-electrolyte. Certain electrolytes, when dissolved in water, do not all fall apart into ions. For example, every tenth molecule does dissociate but 9/10 remain non-dissociated. We call these materials *weak electrolytes*.

Our blood, sweat, and tears (not to mention saliva, urine, and other bodily fluids) are all electrolytes. Given the importance of electrolytes and the fact that the number of particles of an ionic compound, when dissolved in water, multiplies we have to figure out a good way to keep track of the actual concentration of an electrolyte solution.

Let us go back to the basics, to Gibbs free energy. For a crystal of pure salt, under the pressure $p = 101,325$ Pa and at room temperature $T = 298.2$ K, the Gibbs energy is the standard Gibbs energy:

$$G = G_T^\ominus$$

When NaCl is dissolved, for example, in water, its Gibbs energy will be different by a factor containing a logarithm of a property we call activity, a, that is, a measure of "active" NaCl is in the solution:

$$G = G_T^\ominus - RT \ln a$$

There is an old problem with activities – we cannot measure them directly (there is no lab instrument called activity meter) nor can we easily and correctly calculate them. Yet, with electrolytes and the increased number of particles in solution we have to figure out, even approximately, what the activities of the solutes are. So – as is often done in physical chemistry – the problem is split into an easy part and a difficult part. Activity is expressed as a product of concentration, either molar, M, or molal, m, and something called *activity coefficient*, γ. The easy part is of course concentration, M or m, and the difficult part is the activity coefficient, γ.

$$a = m\gamma$$

For lower concentrations we can only hope that γ is close to 1 and this turns out to be the case. This gets considerably more complicated with electrolytes. NaCl in solution dissociates into Na^+ and Cl^- and we need to know activities of both ions. Given that Na^+ is a small positively charged particle and Cl^- a much larger negatively charged particle their activities are expected to be different too. We do not know how to express this by formulas and numbers because, as stated at the beginning of this section, one of the fundamental properties of ions is that they come in pairs. So do their activities and their activity coefficients. We then write the following relation:

$$\gamma(+) \approx \gamma(-) = \gamma_\pm \tag{8-2}$$

Here $\gamma_\pm$ is an "averaged" activity coefficient, which we assume to suit both cations and anions. This is clearly an approximation and we do not really know how good it is. The good news is that, for low-concentration solutions, we can approximately calculate these average coefficients. The theory and formulas for this procedure were developed by Peter Debye and Erich Hückel, Dutch and German physical chemists, respectively, in the early 1920s [1]. According to the Debye–Hückel limiting law, DHLL, approximation, ionic the activity coefficient $\gamma_\pm$ is a function of concentration and charges of all ions in the solution, as follows:

$$\log \gamma_\pm = -A\, z_i^2\, I^{1/2} \tag{8-3}$$

The empirical parameter, A, is a mixture of physical parameters and averaged approximations and usually has the value 0.509. An obvious, but easy to neglect, point is that ionic strength I contains the concentrations and charges of *all ions* present in the solution. Let us do an example.

Problem 8.2 | A strong action requested.

1.2×10^{-3} mol of HCl and 0.9×10^{-3} mol NaOH are completely dissolved and let react in 1 L of water till completion. (A) Write the reaction equation. (B) Calculate the number of moles of products and reactants upon the completion of reaction. (C) Calculate the pH of the solution. Assume 100% dissociation for the acid, base, and salt.

» Solution – Strategy and Calculation

Part (A): As we know from a previous consideration, Fig. 6.3, this reaction goes like this:

$$HCl + NaOH = NaCl + HOH$$

Part (B): The concentration of NaOH is smaller than that of HCl. We say that NaOH is the *limiting reagent* and will react completely to yield NaCl and water. Clearly, some HCl will be left over; let us calculate how much:

$$1.2 \times 10^{-3}\ \text{mol HCl} + 0.9 \times 10^{-3}\ \text{mol NaOH} = 0.9 \times 10^{-3}\text{mol NaCl}$$
$$+\, 0.9 \times 10^{-3}\ \text{mol HOH} + HCl$$

How much HCl is left? All of the HCl did not react with the base simply because there is not enough base, so you have

$$HCl(\text{leftover}) = 1.2 \times 10^{-3} - 0.9 \times 10^{-3} = 0.3 \times 10^{-3}\ \text{mol}$$

Part (C): According to the book, pH is a negative logarithm of the *activity* of H⁺:

$$pH = -\lg a_{H+} = -\lg m(H^+) - \lg \gamma_\pm(H^+)$$

The molality and molarity are almost the same at this low concentration so you can calculate $-\lg m(H^+)$ as

$$-\lg(0.3 \times 10^{-3}) = 3.52$$

And for lg $\gamma_\pm$ (H$^+$) you should use the DHLL approximation:

$$\log \gamma_\pm(H^+) = -A\, z(H^+)^2 I^{1/2}$$

The term $z(H^+)$, the charge of proton, is of course 1, $A = 0.509$, and I, the ionic strength, equals to one-half the sum of the charge concentration product for all ions present in the solution:

$$I = 1/2 \Sigma c_i z_i^2 \tag{8-4}$$

Calculating I is a straightforward exercise; yet it is also easy to make an error as we may not take into account all the present ions. You should first make sure you have listed all ions, their charges, and concentrations, that is

H$^+$	$z^2 = 1$	$c = 0.3 \times 10^{-3}$ m
Cl$^-$	$z^2 = 1$	$c = 1.2 \times 10^{-3}$ m
Na$^+$	$z^2 = 1$	$c = 0.9 \times 10^{-3}$ m

So the ionic strength is then given as

$$I = 0.5 \times \{(0.3 \times 10^{-3})(1)^2 + (1.2 \times 10^{-3})(-1)^2 + (0.9 \times 10^{-3})(1)^2\}$$
$$= 0.5 \times \{2.4 \times 10^{-3}\} = 1.2 \times 10^{-3}$$

And the square root of ionic strength is

$$I^{1/2} = (1.2 \times 10^{-3})^{1/2} = 3.46 \times 10^{-2}$$

And, finally, the activity coefficient for proton in this solution is

$$\log \gamma_\pm(H^+) = -Az(H^+)^2 I^{1/2} = -0.509 \times 1 \times 3.46 \times 10^{-2} = -1.76 \times 10^{-2}$$

Now you use the property that the inverse function of lg is power to ten and write for the activity coefficient

$$\gamma_\pm(H^+) = 10^{[-0.0176]} = 0.960$$

Note that $\gamma_\pm$ (H$^+$) is fairly close to 1 in this solution. You actually do not need $\gamma_\pm$ (H$^+$) to calculate pH, you need its logarithm:

$$pH = -\lg m(H^+) - \lg \gamma_\pm(H^+) = 3.52 - (-1.76 \times 10^{-2}) = 3.54$$

You can see now that, for low concentrations, pH is indeed almost the same as pC$_H$, the negative logarithm of the concentration of H$^+$ ions. Out of habit, we use the pH notation for pC$_H$, that is, the negative logarithm of the *concentration* of H$^+$ ions.

A comment on ionic strength: In solutions of strong acids and bases, with a low over-all concentration of other ions, the effects to the activity coefficient and the pH of the solution are often relatively small and could be ignored. This, however, is not the case in less diluted solutions where the acid–base equilibrium of complex biological molecules can shift significantly with changes in the ionic strength of the solution. The method of pH shifting – by adding salt – is common in biochemical laboratory procedures. Take a look at the following example.

| Problem 8.3 | (Over)Salting a turkey. |

A millimolar solution of tryptophan hydrochloride in acidified water, pH 2.38, reaches an equilibrium to give a 50:50 ratio of cation to zwitterion, Fig. 8.1, right. What is this ratio going to be after you add 0.025 mol of aluminum sulfate to the solution?

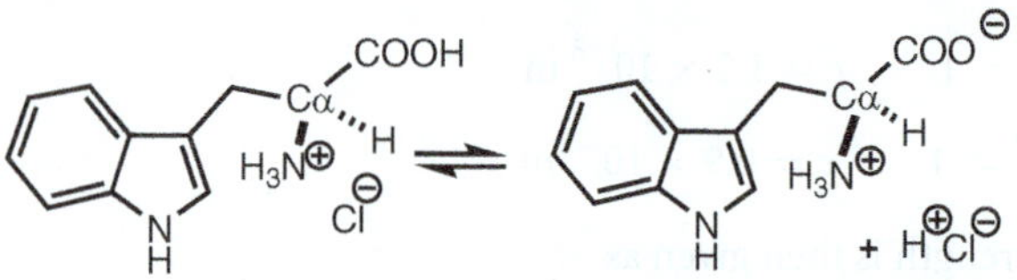

8

■ Fig. 8.1 Tryptophan hydrochloride in equilibrium with its zwitterion

A note on quaternary ammonium salts: Converting tryptophan to tryptophan hydrochloride, $TrpH^+Cl^-$, by addition of hydrochloric acid, HCl, is a common practice in medicinal chemistry and pharmacology. It converts the drugs which contain amino, $-NH_2$, group – of which there are many – into water-soluble salts that can be administered intravenously. Read the fine print on the over-the-counter drug packages and you will see that most of them are in the form of HCl salts.

» Solution – Strategy

This is a problem similar to the one with HCl and NaOH, only with a little twist at the end. The first question asks how much will the pH 2.38 be shifted after addition of 0.025 mol of aluminum sulfate. The second question asks how this change is going to affect the equilibrium between the tryptophan cation and tryptophan zwitterion forms.

What do we start with? Follow the text and break it into separate events:

- First, add 0.001 mol of $TrpH^+Cl^-$ to a water solution at pH 2.38.
- Then add and dissolve 0.025 mol $Al_2(SO_4)_3$.
- You will then determine pH of the solution in the same way you have done for the HCl–NaOH pair in the problem above.
- Finally, insert the new pH value into the $TrpH^+ - Trp^{\pm}$ equilibrium equation to find out how much it has shifted.

Let us start by listing all the ions in the solution:

1. First, we put 1.0×10^{-3} mol ("millimolar") of $TrpH^+Cl^-$ into the solution, 50% of which turns into zwitterions, $Trp^{\pm}$. So you have $0.001/2 = 0.0005$ mol $TrpH^+$ and 0.0005 mol Cl^-. You also have $0.001/2 = 0.0005$ mol of zwitterion, $Trp^{\pm}$. Let us write this down:

   $c(TrpH^+Cl^-) = 0.0005$

   $c(Cl^-) = 0.0005$

   $c(Trp^{\pm}) = 0.0005$

2. There is enough HCl in the solution to maintain pH at 2.38; this will help us calculate the concentration of H^+ and Cl^- ions created by the dissociation of hydrochloric acid. Assuming that at this concentration (millimolar) the ionic activity coefficient is close to one you can calculate the concentration of protons and chloride ions directly from pH according to the following relation: $c(H^+, Cl^-) = 10^{-pH} = 10^{-2.38} = 4.17 \times 10^{-3}$ mol. Let us list these ions one by one:

   $c(H^+) = 4.17 \times 10^{-3}$

   $c(Cl^-) = 4.17 \times 10^{-3}$

3. In step #3 aluminum sulfate dissociates to give 2×0.025 mol Al^{+3} and 3×0.025 mol SO_4^{-2}, so this will amount to the following molar concentrations:

   $c(Al^{+3}) = 0.050$

   $c(SO_4^{2-}) = 0.075$

» Solution – Calculation

You will use the DHLL approximation to calculate the proton activity coefficient:

$$\log Y_{\pm}(H^+) = -Az(H^+)^2 I^{1/2}$$

The term $z(H^+)$, the charge of proton, is of course 1, $A = 0.509$, and I, the ionic strength, equals to one-half the sum of the charge concentration product for all ions present in the solution:

$$I = 1/2\Sigma c_i z_i^2$$

$$I = 0.5 \times \{0.5 \times 10^{-3} \times (\pm)^2(Trp^{\pm}) + 0.5 \times 10^{-3} \times (+1)^2(TrpH^+)$$
$$+ 0.5 \times 10^{-3} \times (-1)^2(Cl^-) + 0.050 \times (+3)^2(Al^{3+}) + 0.075$$
$$\times(-2)^2(SO_4^{2-}) + 4.17 \times 10^{-3} \times (+1)^2(H^+) + 4.17 \times 10^{-3} \times (-1)^2(Cl^-)\}$$

You should for a moment ignore the contribution of free proton and chloride ions as too small and this will give you

$$I = 0.5 \times 0.761 = 0.381$$

Find the square root of I and insert it into the DHLL expression for the logarithm of proton activity coefficient, $\log \gamma_\pm (H^+)$:

$$\log \gamma_\pm(H^+) = -Az(H^+)^2 I^{1/2}$$

$$\log \gamma_\pm(H^+) = -0.509 \times (+1)^2 \times (0.381)^{1/2} = -0.314$$

If you take this number to the power of ten it will give you activity coefficient, $\gamma_\pm (H^+) = 10^{-0.314} = 0.485$, but you do not really need it; insert, instead, the value for $\log \gamma_\pm (H^+)$ directly into the pH equation:

$$pH = -\log m(H^+) - \log \gamma_\pm(H^+)$$

$$pH = -\log(4.17 \times 10^{-3}) - (-0.314) = 2.69$$

So after the addition of aluminum sulfate pH of the solution shifts up, from 2.38 to 2.69. Let us now see how much will the tryptophan acid–base equilibrium change in this new environment. You will use the HH equation to determine the position of the cation–zwitterion equilibrium, as given in Fig. 8.1.

Since only the COOH group dissociates, α-COOH = α-COO$^-$ + H$^+$, the equilibrium expression can be simplified as

$$pK_a = pH - \lg \{a(\alpha\text{-COO}^-)/a(\alpha\text{-COOH})\}$$

You will get the $a(\alpha\text{-COO}^-)/a(\alpha\text{-COOH})$ ratio when you insert the numbers for pK_a and pH and raise the equation to the power of ten:

$$a(\alpha\text{-COO}^-)/a(\alpha\text{-COOH}) = 10^{-(2.38-2.69)} = 10^{+0.31} = 2.06$$

A comment: Now, 2.06 is quite different than 1, wouldn't you say? So you see: addition of aluminum sulfate, an electrolyte chemically unrelated to tryptophan, significantly changes the balance between tryptophan ionic forms. In physical sciences such change is called a *second-order effect*, or second-order phenomenon. The change of pH, following addition of few drops of HCl, is a first-order effect; the salt-induced pH shift is a second-order effect. It also illustrates the significance and a need to know solute *activities*. In many biological systems, with drastically decreased number of solute and solvent molecules, knowing the activities of both solutes and solvents becomes essential to understanding chemical reaction mechanisms at molecular level.

Reference

1. Debye P, Hückel E (1923) Zur Theorie der Elektrolytre. I. Gefrierpunktserniedrigung und verwandte Erscheinung. Phys Z 24:185–206

9 Electrochemistry

Electrochemical reactions are yet another example of how powerful and versatile the *most important equation in physical chemistry* is, stated again

$$\Delta G = \Delta G_T^{\ominus} + RT \times \ln K_{eq}$$

Specifically, it relates an *intrinsic* property – electromotive force, E – to the exchange of electrons and ions. This exchange, whether it is accompanied by making and breaking of chemical bonds or only involves a change in the oxidation state of a metal or other center, is described by the following variant of the above equation:

$$E = E_T^{\ominus} - (RT/zF) \times \ln K_{eq}$$

Here, E and $E_T^{\ominus}$ are equivalents of reaction-induced changes in Gibbs energy and standard Gibbs energy while z and F are electrical normalization factors, the number of exchanged electrons and the charge of the Avogadro's number of electrons, respectively. Much has been written about electrochemical reaction and I will cover it only briefly, using an exemplary and well-known model.

Problem 9.1	The exchange telegraph cell.

Write down the expression for the copper–zinc electrochemical cell. Write the reducing reactions for the half-cells and the redox reaction for the whole cell. Assume that equilibrium has been reached and from the standard Cu^{+2}/Cu and Zn^{+2}/Zn potentials calculate the equilibrium constant.

» Solution – Background

It will be helpful at this point to explain in detail how to make an electrochemical cell and what processes are occurring there. We will start with a classical copper–zinc cell. First, we find a copper block $Cu(s)$ and immerse it into a solution of copper sulfate, $Cu^{2+}(aq) SO_4^{2-}(aq)$, Fig. 9.1. Next, we connect the copper block with a source of electrons, Fig. 9.2. When two electrons, $2e^-$, are supplied to the copper block

P.-P. Ilich, *Selected Problems in Physical Chemistry*,
DOI 10.1007/978-3-642-04327-7_9, © Springer-Verlag Berlin Heidelberg 2010

they combine with a Cu^{2+} ion from the $CuSO_4$ solution to give an atom of solid elementary copper, $Cu(s)$; the copper atom deposits on the copper electrode, Fig. 9.3. If more electrons were supplied to the copper block more $Cu^{2+}(aq)$ ions from the solution will deposit and this will lead to charge imbalance – only $SO_4^{2-}(aq)$ ions will be left – and the process cannot continue spontaneously.

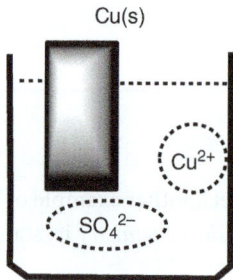

■ **Fig. 9.1**   A solid copper block, $Cu(s)$, in a $Cu^{2+}(aq)$, $SO_4^{2-}(aq)$ solution

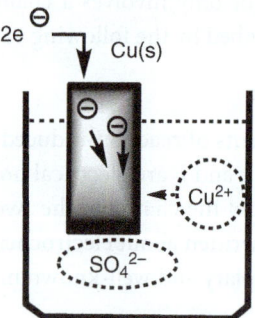

■ **Fig. 9.2**   A $Cu^{2+}(aq)$ ion gets reduced to $Cu(s)$ at the electrode surface

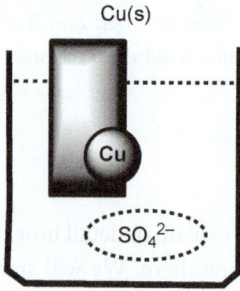

■ **Fig. 9.3** $Cu(s)$ deposited on the copper block

In another experimental setup a zinc block, Zn(s), Fig. 9.4, immersed in an aqueous solution of zinc sulfate, $Zn^{2+}(aq)$ and $SO_4^{2-}(aq)$ spontaneously releases two electrons and enters the solution as $Zn^{2+}(aq)$ ion, Fig. 9.5. The Zn(s) electrode has lost one atom (there is a "hole" in the Zn block), acquires a net charge of two electrons, $2e^-$, and the $ZnSO_4$ solution has a surplus of Zn^{2+} ions, Fig. 9.6.

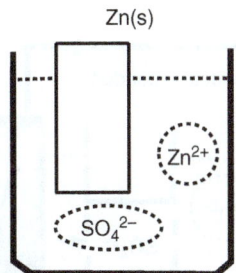

▪ **Fig. 9.4** A solid zinc block, Zn(s), in a $Zn^{2+}(aq)$, $SO_4^{2-}(aq)$ solution

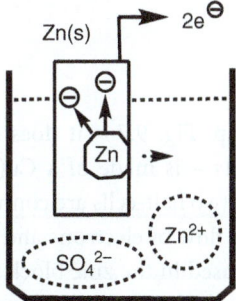

▪ **Fig. 9.5** A Zn(s) atom gets oxidized and turns into $Zn^{2+}(aq)$ ion

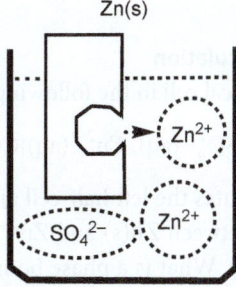

▪ **Fig. 9.6** The Zn(s) block is depleted by one atom

However, this process leads to a charge imbalance – there are more and more Zn^{2+}(aq) ions in the solution while the concentration of the $SO_4{}^{2-}$(aq) ions remains the same – and it cannot continue spontaneously.

The Cu/CuSO$_4$(aq) and Zn/ZnSO$_4$(aq) are so-called half-cells; both have electrochemical potentials but cannot sustain an electrochemical reaction on their own. We have to tie them together.

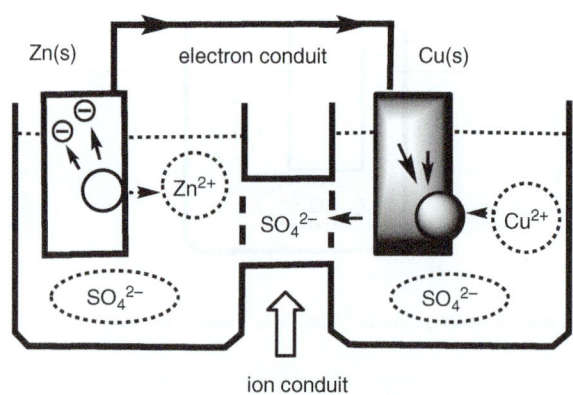

□ Fig. 9.7 A scheme of Daniell cell

So our next experimental setup, Fig. 9.7 – it does look a little complicated but we already know its main parts – is made of a Cu(s)|CuSO$_4$(aq) half-cell and a Zn(s)|ZnSO$_4$(aq) half-cell. The two half-cells are connected in two ways: by wire (or electron conduit) which allows flow of electrons and by a salt bridge which allows ion transfer. The electrons released in the zinc block flow through the wire toward the copper block and reduce Cu^{2+} ions which get deposited. The surplus of negative $SO_4{}^{2-}$ anions from the copper half-cell travels toward the zinc chamber to make up for the charge balance with the surplus of Zn^{2+} ions there. Everything works the way it should and we have a fully operational cell – a Daniell cell.

» Solution – Strategy and Calculation
We represent this electrochemical cell in the following way:

$$Zn(s)|Zn^{2+}(aq), SO_4^{2-}(aq)||SO_4^{2-}(aq), Cu^{2+}(aq)|Cu(s) \qquad (9\text{-}1)$$

Here, the Zn/Zn^{2+} pair constitutes the left half-cell and the Cu/Cu^{2+} pair the right half-cell. The vertical bar, as between Zn(s) and Zn^{2+}(aq) and Cu^{2+} (aq) and Cu(s), is a symbol for *phase boundary*. What is a phase boundary? It is the place – point, line, surface – where materials of different phases meet. Think of a glass beaker filled with water: the glass wall is a phase boundary. The double vertical line, ||, represents a salt bridge in electrochemical cells, that is, a conduit for ions.

We first write the *reductive reaction* for the left side of the cell:

$$Zn^{2+} + 2e^- \rightarrow Zn(s) \tag{9-2}$$

The reductive reaction for the right side is written in a similar way:

$$Cu^{2+} + 2e^- \rightarrow Cu(s) \tag{9-3}$$

The overall equation for the two half-cells is written as the right reduction reaction minus the left reduction reaction:

$$Cu^{2+}(aq) + 2e^- - Zn^{2+}(aq) - 2e^- = Cu(s) - Zn(s)$$

After you reshuffle the negative terms across the = sign and cancel the electrons you will have the cell reaction:

$$Cu^{2+}(aq) + Zn(s) \rightarrow Cu(s) - Zn^{2+}(aq) \tag{9-4}$$

As the copper and zinc ions will have different chemical potentials there has to be some "process" going on after we made these connections; we describe it by the Gibbs free energy change or the change in chemical potentials:

$$\Delta G = \Delta G_T^\ominus + RT \ln Q \tag{9-5}$$

where Q, as before, is a ratio of the activities (concentrations, pressures) of the reaction products and reactants:

$$Q = a(Zn^{2+})\, a(Cu(s)) / a(Cu^{2+})\, a(Zn(s))$$

Since the activities of solid Cu and solid Zn are taken to be one, Q becomes the ratio of the activities of Zn^{2+} and Cu^{2+} ions in the two solutions. Notice that no electrons figure out in the final equation and no work needed for an electron to traverse the Zn block is included. Similarly, you do not have to include the work needed to unite two electrons, $2e^-$, in the copper electrode with a Cu^{2+} ion in solution. These contributions – electron motion and electron–ion interaction – are included in the *electrochemical potential, E*, of the half-cells. The electrochemical potential for the complete cell is defined by the following relation:

$$\Delta G^\ominus = -z F E \tag{9-6}$$

Here z is the number of electrons and F is the so-called Faraday constant, given as one mole of elementary charges:

$$F = (1\ \text{elementary charge}) \times \text{Avogadro constant}$$
$$F = 1.602176476 \times 10^{-19}\ [C] \times 6.02214179 \times 10^{+23}\ [mol^{-1}]$$
$$= 96{,}485.3398\ [C\,mol^{-1}]$$

You will write similar expression for the standard free energy:

$$\Delta G^\ominus = -z F E_T^\ominus \tag{9-7}$$

T stands for the ambient temperature, 298.15 K, if not specified otherwise. You may express the ratio Q and eventually the activities (and concentrations) of the metal ions in solution using the experimental values for the cell potential. But, before you do this let me make an important comment.

A comment on electrochemical potentials: The Gibbs free energy of an electrochemical cell is a *cumulative*, or *extensive*, or *extrinsic* property, that is, the larger the cell the greater the $\Delta G^\ominus$ value. The electrochemical potential, E, however, is a *specific*, or *intensive*, or *intrinsic* property as it does not depend on the size of the electrochemical cell.

Now proceed by dividing (9-5) by $-z\,F$ and replacing the ratio $-G/zF$ by E. This is the so-called *Nernst's* equation, after the German physical chemist Walter Nernst:

$$E = E^\ominus - (RT/zF)\ln Q \tag{9-8}$$

The value for $E^\ominus$ is obtained from the tables of experimental standard potentials for half-cells:

$$E_L{}^\ominus(Zn^{2+}/Zn(s)) \text{ is } -0.76 \quad and \quad E_R{}^\ominus(Cu^{2+}/Cu(s)) + 0.34\,V$$

The cell standard potential is obtained when you subtract the left standard potential from the right standard potential:

$$E^\ominus = E_R{}^\ominus - E_L{}^\ominus = +0.34\,V - (-0.76\,V) = +1.10\,V$$

The Nernst's equation will now read

$$E = 1.10 - (RT/zF)\ln Q$$

Since at equilibrium there is no net change in the Gibbs free energy, $\Delta G^\ominus = -zFE = 0$, the ratio Q equals the equilibrium constant, K_{eq}, and is proportional to the standard potential:

$$E^\ominus = +(RT/zF)\ln K_{eq} \tag{9-9}$$

From here we obtain K_{eq}, the ratio of the activities of Zn^{+2} (numerator) and Cu^{+2} (denominator) ions:

$$\ln K_{eq} = 2 \times 96{,}485\,[C\,mol^{-1}] \times 1.10\,[V]/8.31\,[J\,K^{-1}\,mol^{-1}] \times 298.15\,[K]$$

$$\ln K_{eq} = 85.673 \text{ and}$$

$$K_{eq} = a(Zn^{2+})/a(Cu^{2+}) = \exp(\ln K_{eq}) = e^{85.673} = 1.61 \times 10^{+37}$$

A comment on electrochemical cells: This large, practically infinite ratio of the activity of Zn^{2+} ions to the activity of Cu^{2+} ions is telling us that in a cell designed as the one given in Fig. 9.7 there is a very strong thermodynamic drive to convert $Zn(s)$ atom into Zn^{2+} ions (oxidation) and to convert Cu^{2+} ions into solid copper, $Cu(s)$

(reduction). A practical consequence of these electrochemical processes in the cell is a flow of electrons through the external wire. The cell is a *battery*, a so-called *galvanic* cell. A galvanic cell of this design is known as *Daniell cell*. In a modified form, known as *crowfoot* cell, it was the major power source for early telegraphy in the 19th century. Daniell cell illustrates the basic principles of other types of electrochemical cells. And no – your cell phone is not powered by a Daniell cell.

Opposite from galvanic cell is an *electrolytic* cell where external electric current is applied in order to induce chemical changes within a cell; a textbook example is a cell used to electrolyze liquid water into gaseous H_2 and O_2. Electrolytic cells have a number of important industrial applications and are the subject of technical-engineering electrochemistry.

Make a note: *galvanic cells* (batteries) and *electrolytic cells.*

A note on half-cell potential values: The electrode (or the half-cell) potentials in the example above, –0.76 V for the Zn^{2+}/Zn and +0.34 V for Cu^{2+}/Cu, are not some numbers that we know as a fact of nature but *relative* potentials determined by comparison to the electrochemical potential of the so-called standard hydrogen electrode, SHE,

$$Pt(s)|H_2(g)|H^+ (aq)$$

A SHE is a platinum wire immersed in acidified aqueous solution (HCl) and surrounded by a tube through which bubbles dihydrogen gas at standard pressure. The reduction reaction and the Nernst equations for this half-cell are given as

$$H^+ + e^- \rightarrow 1/2\ H_2(g) \tag{9-10}$$

$$E = E^\ominus - (RT/1\ F) \ln \{p(H_2)^{1/2}/a(H^+)\} \tag{9-11}$$

With the standard potential set at zero, $E^\ominus$ (SHE) = 0.0 V, and the pressure (the so-called *fugacity* – or "adjusted" pressure) being one, the SHE potential, expressed in volts, reads

$$E = 0.0 - 5.916 \times 10^{-2}[V] \times pH \tag{9-12}$$

Here, we have used the well-known relation for conversion of two logarithms with different bases, e and 10:

$$lg_a(x) = lg_a(b) \times lg_b(x), \text{ that is, } lg_e(x) = lg_e(10) \times lg_{10}(x)$$

Or, using the common way the $lg_e()$ and $lg_{10}()$ are written we have the following relation:

$$\ln a(H^+) = \ln(10) \times lg\, a(H^+)$$

The expression $(RT/F) \times \ln(10)$ is 59.16 mV and is a useful constant in calculations of electrochemical potentials that are pH dependent. The expression (9-12) also tells us an important fact: the SHE potential will change with pH in the half-cell. At physiological conditions, $pH = 7$, the SHE potential is a significant negative number, that is

$$E(\text{pH } 7) = 0.0 - 5.916 \times 10^{-2} \, [\text{V}] \times 7 = -0.414 \, \text{V}$$

We also need *extra* energy, $\Delta G = -(-0.414 \, [\text{V}] \times 96 \, 485 \, [\text{C mol}^{-1}]) \approx 40 \, \text{kJ}$, to reduce a mole of protons under these conditions. Now that the standard changed everything else has changed. For example, the standard potential of the very important reaction of reduction of dioxygen to water, $E^{\ominus}(O_2/O_2^{4-})$ is $+1.229$ V relative to the standard SHE and $1.229 - 0.414 = 0.815$ V when compared to the "physiological" (or biochemical) standard hydrogen electrode, SHE (pH 7). The "take home message" from this note is this: When comparing two half-potentials they have to be determined relative to the standard hydrogen electrode at the same pH.

Make a note: E (SHE) $= 0.0$ V and E (SHE, pH 7) $= -0.414$ V.

9.1 Biological Electrochemistry

Electrochemical cells in living organisms are based on the same principle as Daniell cell but are of different design (i.e., no little copper blocks lodged in your blood vessels). Creation, transfer, and consumption of electrons and ions occur in every biological-chemical system and are the essential prerequisites of life. The ultimate and most important electrochemical reaction in living world is *respiration*. Earlier, when talking about gases, we described respiration as inhalation of oxygen and exhalation of carbon dioxide. On a molecular scale, respiration is a cascade of rather complex reduction–oxidation (redox) reactions which end up with a terminal electron acceptor. We, humans, and many other organisms use O_2 as the electron acceptor but this does not have to be the case; the commensal bacteria in our intestines (commensal = sharing of food) or the deep soil and deep ocean bacteria use other elements, like sulfur or even arsenic, as terminal electron acceptors. Another example of bio-redox reactions are processes involved in the response of our bioelectrochemical agents to xenobiotics – chemicals not normally introduced or expected to be introduced into organism. Furthermore, in living organisms, the membranes surrounding subcellular parts (mitochondria), cells, or even whole bodies (skin) are semi-permeable, that is, they restrict molecular traffic either to one direction or to one size/type of particles. The exchange of fluids and ions through a semi-permeable membrane is referred to as dialysis. The separation and removal of metabolic products and toxins, taking place in kidneys (natural or artificial), is based on dialysis. The presence of ions of different size and charge, on both sides of a semi-membrane, results in a material imbalance and an emergence of electrical potential across the membrane. Though small in absolute values, $10^{-2}–10^{-3}$ V, this electric potential is a vital sign of living organisms. Such is *Donnan* potential, named in honor of the Irish physical chemist George Donnan. We will do one example of each: an oxidoreduction of xenobiotics, a redox step in a respiratory chain, and a membrane potential.

Problem 9.2 | Being good to your liver (Don't drink and drive)

The following redox (*red*uction– *ox*idation) reaction, Fig. 9.8, illustrates what happens in your liver after you had a drink: The C1 in the ethanol from the drink, CH_3CH_2OH, gets *oxidized* by two electrons (i.e., it loses $2e^-$), to give acetaldehyde, CH_3CHO. This happens through the action of enzyme LADH, liver alcohol dehydrogenase, found in liver and other tissues in our body [1–3]. An enzyme – as you probably know – is a large peptide molecule, a biological catalyst which makes this reaction possible. But not all of the protein is involved (at least not directly) in the reaction given above. The electrochemical reaction is taking place at the $NADP^+$/NADPH redox center in the enzyme and such center in an enzyme is called the *active site.* In the process, the C4 in the pyrimidine ring of the nicotineamide adenine dinucleotide phosphate center in LADH gets *reduced* by two electrons, i.e., it accepts $2e^-$. Given that the standard reduction potential of nicotineamide adenine dinucleotide phosphate, $NADP^+$, $E^{\ominus}(NADP^+/NADPH)$ at pH = 7 is −0.320 V, and that of acetaldehyde/ethanol pair, at the same pH, $E^{\ominus}(CH_3CHO/CH_3CH_2OH)$ is −0.197 V, estimate the Gibbs energy for this reaction. Determine whether the reaction is thermodynamically spontaneous at pH 9.9 – the optimal pH for this reaction.

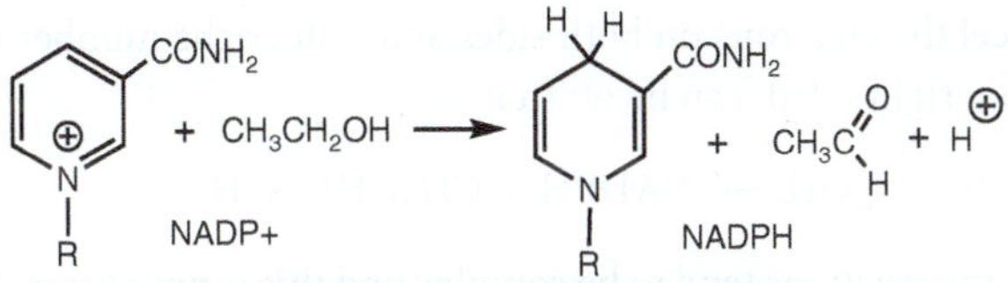

Fig. 9.8 $NADP^+$, in a reaction with ethanol gets reduced to NADPH while oxidizing the alcohol to aldehyde

Solution – Strategy and Calculation

What is this problem about? Not exactly a Daniel cell but – on a closer view – not very different from it either: ethanol gives away two electrons so ethanol is our zinc electrode. Then the rather complicated $NADP^+$ molecule (of which we have shown only the 1/7th part) must be the copper electrode. So there – problem solved.

$NADP^+$/NADPH is the major redox molecular pair in living organisms. Remember this symbol: $NADP^+$ and NADPH. The oxidoreductive reaction given in the scheme above is a relatively simple reaction which, like the Daniell cell, has the left and the right side. The reductive reaction for the left half-cell goes like this:

LEFT reaction: $CH_3CHO + 2e^- + 2H^+ = CH_3CH_2OH$

How do we know that two – and not one or three – electrons are needed to reduce a molecule of acetaldehyde to ethanol? Take a look at the structural formulas of acetaldehyde and ethanol, Fig. 9.9. What do you see? The electrochemical reaction occurs on C1. It changes its formal oxidation state, $FOx(C)$, from +1 in acetaldehyde to –1 in ethanol, that is, by *two* electrons. For a reminder how to assign a formal oxidation state to carbon atom take a look at Fig. 9.9. A +1 formal charge is assigned to each C–H bond, a –1 charge to carbon atom connected to O, N, Cl, or other electronegative atom. Finally, a zero is assigned to a C–C bond.

RIGHT reaction: $NADP^+ + 2e^- + H^+ = NADPH$

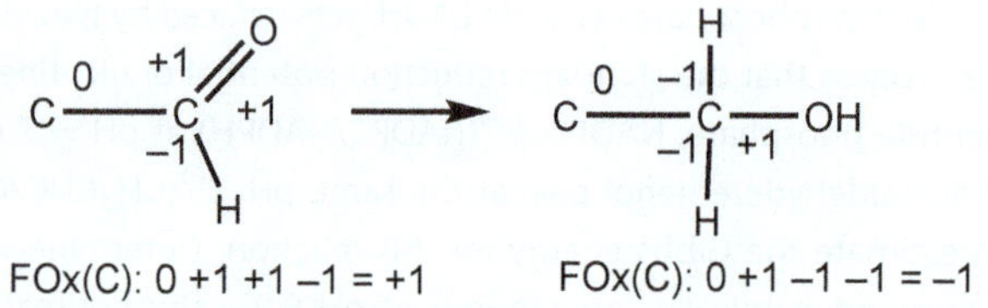

FOx(C): 0 +1 +1 −1 = +1   FOx(C): 0 +1 −1 −1 = −1

Fig. 9.9 The change in formal oxidation state on C1 when ethanal gets reduced to ethanol

When we cancel the electrons on both sides and reduce the number of protons the overall reaction, right – left, can be written as

$$NADP^+ + CH_3CH_2OH \rightarrow NADPH + CH_3CHO + H^+$$

Note that enzyme reactions tend to be complex and this is not necessarily a stoichiometric relation; however, we can write a formal equation for the change of potential (or *electromotive force*, EMF):

$$E = E^\ominus - (RT/zF)\ln Q$$

The reaction quotient, Q, is given as

$$Q = \{a(NADPH)\, a(CH_3CHO)\, a(H^+)/a(NADP^+)\, a(CH_3CH_2OH)\}$$

The standard Gibbs energy for this reaction is obtained when we subtract the Gibbs energy of the left reaction from that of the right reaction:

$$E^\ominus = E^\ominus(RIGHT) - E^\ominus(LEFT)$$

$$E^\ominus = -0.320\,[V] - (-0.197)\,[V] = -0.123\,V$$

What can we do now? How do we evaluate the change in the total Gibbs energy of the reaction when we don't know anything about the activities or even concentrations of the reactants and products? True – but we can also use a common sense and, perhaps, find a way out of this problem. First – the NADP$^+$. We do not know its activity – and it will be some time before we are able to measure this kind of physical chemical properties in enzymes – but we do know that in the enzyme active site there are as many NADP$^+$ molecules as there are NADPH molecules (one, per each of the two active sites in LADH). So the *ratio* of activities of NADP$^+$ and NADPH is one. The same goes for ethanol: one molecule of ethanol in the enzyme active site will get converted to one molecule of acetaldehyde. So the ratio of the activities of acetaldehyde and ethanol is about one.

This leaves us with $\ln a(H^+)$; $Q \approx (1)\times(1)\times \ln a(H^+)$. We do not know the value of $\ln a(H^+)$ but we do know pH, that is, $-\lg_{10} a(H^+)$. All you have to do now is to convert natural logarithms of $a(H^+)$ to logarithm base ten of and multiply it by -1. This will give you the following equation:

$$E = E^{\ominus}+(0.0592/z) \times \text{pH}$$

For the given pH, reportedly optimal for oxidation of ethanol to acetaldehyde, you will have for the reaction potential

$$E = E^{\ominus}+0.0592/2 \times 9.9 = -0.123\,[V] + 0.293\,[V] = +0.170\,V$$

And the standard Gibbs free energy is

$$\Delta G = -zFE^{\ominus} = -2 \times 0.170[V] \times 96{,}485[Cmol^{-1}] = -3.281 \times 10^4\,V\,C\,mol^{-1}$$
$$= -32.8\,kJ\,mol^{-1}$$

So – the reaction is spontaneous. The ethanol you have consumed at that party last night is turning in your liver into acetaldehyde, initiating a major chemical poisoning better known as – hangover. This is bad enough but not as bad as if you, by some accident, have consumed methanol, instead of ethanol. LADH will turn methanol into formaldehyde which with the help of another enzyme gets further oxidized into formic acid. The physiological response to larger amounts of formic acids in a body are not hangover but – blindness and death. Fortunately, methanol reacts very slowly with LADH – we say it is a very poor substrate – and a simplest first aid to a person poisoned by methanol is a large dose of ethanol (diluted by water, of course). So – now you know that a strong drink could be life-saving.

Problem 9.3	An electrochemical sequel – the fate of the NADPH/NADP$^+$ brothers.

In order to restart biological oxidation NADPH (or the very similar NADH – nicotineamide adenine dinucleotide) has to be converted back to its oxidized form NADP$^+$ (NAD$^+$) by releasing two electrons to an electron acceptor. In humans and many other organisms the terminal 2-electron acceptor is the gas we inhale, dioxygen, as given by the following half-cell reaction:

$$1/2\,O_2(g) + 2e - +2H^+ \rightarrow H_2O(aq) \qquad E = +0.815\,V(pH7)$$

The energy created by reduction of dioxygen, or the respiration process, is divided into many parts and used to fuel several energy-needing (endergonic) reactions in our body, like the very important process of oxidative phosphorylation. (A) Find out how much energy is released by reduction of dioxygen to water per mole of NADPH. (B) Compare this energy to the energy created in anaerobic bacteria that use sulfite, SO_4^{2-}, as the terminal electron acceptor, to re-oxidize NADPH (NADH) in a reaction chain given by the following (very simplified) scheme:

$$HSO_3^- + 6e^- + 6H^+ \rightarrow HS^- + 3H_2O \qquad E^{\ominus} + 0.301\,V\,[4]$$

9

» Solution (A) – Strategy and Calculation

The strategy we need here is simple and straightforward: we are going to calculate the difference in the standard potentials between the two half-cells, $NADP^+/NADPH$ and O_2/H_2O. Next, using Nernst's equation, we will calculate the change in the standard Gibbs free energy of this system. Done.

The left half-cell is the one supplying electrons to the process: NADPH. Yes, the reduced form of nicotineamide adenine dinucleotide phosphate is the zinc electrode in the Daniell cell, and we re-write its reductive half-reaction:

$$\text{Left: } NADP^+ + 2e^- + H^+ = NADPH \qquad E^{\ominus} = -0.320\,V(pH7)$$

Dioxygen, like the copper electrode in Daniell cell, receives the two electrons and is the right end of the electrochemical cell; we re-write its reductive half-reaction as follows:

$$\text{Right: } 1/2\,O_2(g) + 2e^- + 2H^+ \rightarrow H_2O \qquad E^{\ominus} = 0.815\,V(pH7)$$

When the left part is subtracted from the right part of the reaction, the standard potential and the change in the Gibbs standard energy for this electrochemical pair are given in short order as

$$1/2\,O_2(g) + NADPH + H^+ \rightarrow H_2O + NADPH^+$$
$$E^{\ominus} = +0.815\,[V] - (-0.320)\,[V] = +1.135\,V$$

$$\Delta G^{\ominus} = -z\,E^{\ominus}\,F = -2 \times 1.135\,[V] \times 96{,}485\,[C\,mol^{-1}] = -2.190 \times 10^5\,V\,C\,mol^{-1}$$
$$= -219\,kJ\,mol^{-1}$$

This is almost 220 kJ per mole of dioxygen – quite an energy punch and – it is used to drive a lot of important little gadgets in our body.

Solution (B) – Calculation

For the NADPH–sulfate ion pair we have the following relations and numbers:

Right: $SO_3^- + 6e^- + 6H^+ \rightarrow SH^- + 3H_2O$ $E^\ominus = 0.301\ V\ (pH\ 7)$

Left: $NADP^+ + 2e^- + H^+ = NADPH$ $E^\ominus = -0.320\ V\ (pH\ 7)$

And the overall budget gives us

$SO_3^- + 3NADPH + 6H^+ \rightarrow SH^- + 3H_2O + 3NADP^+$
$E^\ominus = +0.301\ [V] - (-0.320)\ [V] = +0.621\ V$

$\Delta G^\ominus = -zE^\ominus F = -6 \times 0.621\ [V] \times 96{,}485\ [C\ mol^{-1}] = -3.595 \times 10^5\ V\,C\,mol^{-1}$
$= -360\ kJ\ mol^{-1}$

So it is -360 kJ for sulfate vs. -219 kJ for dioxygen. It appears it is more profitable energy-wise to be spending life as a little bacteria, buried in an airless city sewage tank and using sulfate for respiration, than to live outside and enjoy fresh air. What do you think?

Problem 9.4 | **It's all electrical (A little kidney in your Erlenmeyer flask).**

A 0.0015 M solution of charged peptide, $Z_M = +10$, is dialyzed against a 0.100 M CsCl solution. (A) Calculate the ratio of concentrations of Cl^- ion inside and outside the dialysis cell and the resulting Donnan potential. (B) In the second experiment the polypeptide concentration is changed and the Donnan potential now reads 0.0217 V. What is the polypeptide concentration?

Solution (A)– Strategy

It may be easier to understand this riddle if we break the experiment into three stages and three schemes (Figs. 9.10, 9.11 and 9.12):

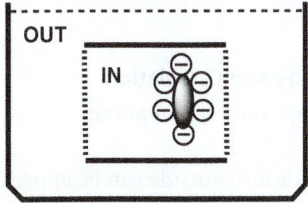

■ **Fig. 9.10** **(Stage 1)** There are two chambers: the inside, IN, and the outside, OUT, chamber. They are separated by a semi-permeable membrane. We fill the inside chamber with a solution of a highly charged peptide, $Z_P = +10$. The concentration of the peptide is $c_P(IN) = 0.0015\ mol\ L^{-1}$. The counterion is chloride, Cl^-. As the charge of Cl^- is 10-fold lower, $Z_{Cl} = -1$, its concentration must be 10 times that of peptide, $c_{Cl}(IN) = 0.0015 \times 10 = 0.015\ mol\ L^{-1}$, to maintain electrical neutrality

The Cs^+ and Cl^- ions can pass the membrane but the peptide cannot (Fig. 9.11). There are more ions outside and their Gibbs energy is higher than the Gibbs energy of ions inside, $G(OUT) > G(IN)$.

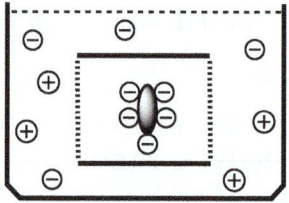

9

■ **Fig. 9.11 (Stage 2):** The outside chamber is filled by electrolyte, Cs^+Cl^-. The ion charges are equal, $Z_{Cs} = Z_{Cl} = |1|$, and the concentration is $0.100\ mol\ L^{-1}$

A certain amount of Cs^+ and Cl^-, x mol L^{-1}, from outside enter the inside chamber and an equilibrium is reached; at that point there is no more change of Gibbs energy: $\Delta G = 0$.
The concentration of Cs^+ inside now equals x: $c_{Cs}(IN) = x$ mol L^{-1}.

The concentration of Cl^- ions inside has increased by x, and we write for the concentration of Cl inside the chamber: $c_{Cl}(IN) = 10 \times 0.0015 + x$ mol L^{-1}.

At equilibrium, the activities of ions inside and outside are the same: $a(IN) = a(OUT)$ (Fig. 9.12).

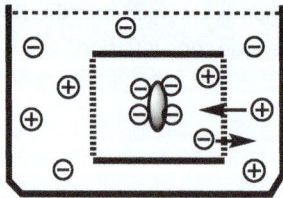

■ **Fig. 9.12   (Stage 3)** The concentration of CsCl outside has decreased by x, and we write the concentration of CsCl outside the chamber as $c_{CsCl}(OUT) = 0.1 - x$ mol L^{-1}

≫ Solution A– More Strategy and Calculation
Now we have to make in important *assumption*:

- The activities of ions inside and outside can be approximated by their concentrations.

Given that the concentrations are relatively low this is likely a reasonably good approximation for Cs^+ and Cl^- ions. However, given the high charge, $Z = +10$, this may be a questionable assumption as far as the charged peptide goes. We cannot assess the possible error without additional experimental data and will proceed as usual, keeping our eyes and minds open for a possible need for a correction. This approximation allows you to write the following, very useful expression:

$$c_{Cl}z_{Cl}\,(IN) \times c_{Cs}z_{Cs}\,(IN) = c_{Cl}z_{Cl}\,(OUT) \times c_{Cs}z_{Cs}\,(OUT)$$

Dropping the charges as being all 1, this expression reads

$$c_{Cl}\,(IN) \times c_{Cs}\,(IN) = c_{Cl}\,(OUT) \times c_{Cs}\,(OUT)$$

Insert the expressions for concentrations, in mol L^{-1}, derived in Stage 3, and you will get

$$(0.0015 \times 10 + x) \times x = (0.100 - x)(0.100 - x)$$

Now solve the equation for x, the amount of Cs^+Cl^- that passed from outside to the inside chamber:

$$x = (0.100\,\text{mol}\,L^{-1})^2/(0.015\,\text{mol}\,L^{-1} + 2 \times 0.100\,\text{mol}\,L^{-1}) = 4.65 \times 10^{-2}\,\text{mol}\,L^{-1}$$

So the equilibrium concentration of Cl^- ions inside and outside is now

$$c_{Cl}\,(IN) + x = 0.146\,\text{mol}\,L^{-1} \text{ and } c_{Cl}\,(OUT) = 0.100 - 4.65 \times 10^{-2}$$
$$= 5.35 \times 10^{-2}\,\text{mol}\,L^{-1}$$

There is clearly an imbalance of the individual ion concentrations inside and outside and an electrical membrane potential develops, proportional to the difference of concentrations (activities) of Cl^- ions inside and outside:

$$E_D = -(RT/zF) \times \ln a(Cl, OUT)/a(Cl, IN) \approx -(RT/zF) \times \ln c_{Cl}\,(OUT)/c_{Cl}\,(IN)$$

When you insert the values for Cl^- concentrations we just calculated you will get for Donnan potential, E_D

$$E_D = -\{8.314\,[J\,K^{-1}\,mol^{-1}] \times 298.15\,[K]/1 \times 96,485\,[C\,mol^{-1}]\}$$
$$\times \ln\,(0.0535/0.1465)$$

$$E_D = -2.569 \times 10^{-2}\,[J\,C^{-1}] \times (-1.007) = +2.59 \times 10^{-2}\,[J\,C^{-1}]$$
$$= 0.026\,V$$

In case you wonder where this unit V, volt, has come from, recall that joule = coulomb × volt, that is, $J = C \times V$.

» Solution B– Hints and Suggestions

In order to find an answer to the second part of the riddle you should keep in mind the following:

- The Donnan potential in the second experiment has decreased, from 2.59×10^{-2} V to 2.17×10^{-2} V.

- The drop in Donnan potential is caused by the increased *ratio* of concentration of the peptide counterion, Cl^-, outside and inside the cell.
- Since the initial concentration of Cl^- ions outside the cell is unchanged, $c_{Cl}(OUT) = 0.100$ mol L^{-1}, this means that the concentration of the peptide is *lower* in the second experiment. Its charge is of course the same, $Z_P = +10$.
- The initial concentration of peptide in the second experiment will be $c_2(P)$ and the initial concentration of the Cl^- counterion, inside the chamber will be, $c_{2Cl}(IN) = c_2 \times 10$. You need to know what c_2 is.
- From the value for the Donnan potential and the equation $E_D = -\{RT/zF\} \times \ln r$ you will be able to calculate r, the ratio of the *equilibrium* concentrations of $Cl^-(OUT)$ and $Cl^-(IN)$. Raise the whole expression to exp[] and solve for r.
- This is like in the first experiment: When you put the inside chamber in contact with the outside chamber, filled with 0.1 mol L^{-1} CsCl, the Cs^+ and Cl^- ions, being of higher concentration, will travel from outside to inside. How much? Again, you don't know, so you may write y mol L^{-1}.
- Now where do you stand? There is c_2 which you need to know and there is y which you do not know: two unknowns. I suggest you try these two equations: $r = (c_{Cl}(OUT) - y)/(c_2 \times 10 + y)$ and $y = [c_{Cl}(OUT)]^2/(c_2 \times 10 + 2 \times c_{Cl}(OUT))$. Make a good substitution and solve for $c_2 \times 10$. This will probably be a quadratic equation. One solution will be good. $c_{Cl}(OUT)$ is of course the initial concentration of CsCl in the outside compartment: 0.100 mol L^{-1}.

Now – this will keep you busy for some time. It is not too difficult a problem, you only have to do a careful bookkeeping of all the terms. When done, you may want to compare the result for c_2 with the one I have obtained: $c_2(peptide) \approx 0.005$ mol L^{-1}.

References

NADP$^+$/NADPH

1. Wermuth B, Münch JD, Wartburg JP (1977) Purification and properties of NADPH-dependent aldehyde reductase from human liver. J Biol Chem 252:3821–3828
2. Lange LH, Sytkowski J, Vallee BL (1976) Human liver alcohol dehydrogenase: purification, composition, and catalytic features. Biochemistry 15:4687–4693
3. Ferguson-Miller S, Babcock GT, Yocum C (2007) Photosynthesis and respiration. In: Bertini I, Gray HB, Stiefel EI, Valentine JS (eds) Biological inorganic chemistry. Structure & reactivity. University Science Books, Sausalito
4. Xavier AV, LeGall J (2007) Sulfur metabolism. In: Bertini I, Gray HB, Stiefel EI, Valentine JS (eds) Biological inorganic chemistry. Structure & Reactivity. University Science Books, Sausalito

Table IV

Summary of electrochemistry – what have we learned in this section?

Review the material we have covered, write down the new words, and then re-write them in symbols and numbers.

Words and Phrases	Symbols, Formulas and Numbers
Dissociation of electrolytes	$NaCl + H_2O \rightarrow Na^+(aq) + Cl^-(aq)$; two ions
Ion activity	$Al_2(SO_4)_3 + H_2O \rightarrow 2Al^{3+}(aq) + 3SO_4^{2-}(aq)$; five ions $(Na^+) = \gamma(+)\,(Na^+) \times c(Na^+)$
Activity coefficients	$\gamma(+) \approx \gamma(-) = \gamma_\pm$
Debye–Hückel limiting law	$\log \gamma_\pm\,(H^+) = -A\,z^2\,I^{1/2}$
Ionic strength of a solution	$I = \frac{1}{2}\,\Sigma\,c_i\,z_i^2 = \frac{1}{2}(c(Na^+)(+1)^2 + c(Cl^-)(-1)^2)$
Colligative properties: freezing point depression	$\Delta T_f = K_f \times$ (# moles of solute/1,000 g solvent)
Daniell electrochemical cell	$Zn_{(s)}|Zn^{+2}{}_{(aq)}, SO_4^{-2}{}_{(aq)}||SO_4^{-2}{}_{(aq)}, Cu^{+2}(aq)|Cu_{(s)}$
Electromotive force	$\Delta G = -z\,E\,F$
Nernst equation	$\Delta G = \Delta G^\ominus + (RT/zF) \times \ln Q$
Standard potentials	$-3.1\,[V] \leq E^\ominus \leq +2.9\,[V]$
Reduction potential	$E(Cu^{+2}/Cu_{(s)})$: $Cu^{+2} + 2e^- \rightarrow Cu(s)$
Electrochemical cell potential	$E(cell) = E(right\ half\text{-}cell) - E(left\ half\text{-}cell)$
An example of biological oxidoreduction	$2\,Myoglobin(Fe^{+3}) + NADPH \rightarrow 2\,Mb(Fe^{+2}) + NADP^+$
Donnan effect and potential	$E_D = -\,(RT/zF)\ln\{[Cl^-, out]/[Cl^-, in]\}$, out, in = outside, inside semi-permeable membrane; Cl^- is peptide counterion

Part V

Kinetics

10 Kinetics 131

10   Kinetics

Kinetics, as you probably know, is about the rate of change. As the original concept and word –*kinesis* in old Greek meant movement or motion – kinetics applies to chemical reactions and also many other types of changes, for example, changes in biological processes, physical changes, or the changes in the amount of money on your bank account. Kinetics is also about observing and recording the speed of change rather than trying to understand its causes (that comes later, when you get really hooked on physical chemistry). Let us look at this kitchen experiment.

» Example
Making a tea

Turn on a gas stove, fill a pot with water, measure the water temperature with a lab thermometer ($t = 15°C$), place the pot on the stove, and look at your watch. After 5 min the temperature of the water reads 55°C, after another 5 min – it will be simpler if you keep the time intervals equal – the temperature is 75°C, then 85°C, then 90°C, then 92.5°C and now you can stop your experiment. Draw these temperatures as equally spaced bars and you get a simple diagram, like the one in Fig. 10.1 .

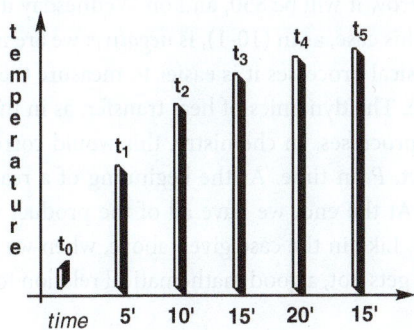

■ Fig. 10.1   Increase of water temperature with time

P-P. Ilich, *Selected Problems in Physical Chemistry*,
DOI 10.1007/978-3-642-04327-7_10, © Springer-Verlag Berlin Heidelberg 2010

If you connect the tops of bars you will get a curve that shows a growth – a growth slowing down with time. Mathematicians have a name for this curve – they call it an exponential saturation curve: $y(x) = 1 - \exp[-kx]$.

As the curve in Fig. 10.1 shows temperature growth slows down with time. But the *rate* of growth remains the same. What is the rate of growth? Since you are looking at a change you will subtract first temperature from second temperature, then subtract second from third, and so on, to get a series: $\Delta t_1 = 55 - 15 = 40$, $\Delta t_2 = 75 - 55 = 20$, $\Delta t_3 = 85 - 75 = 10$. Now divide Δt_2 by Δt_1 and you get $20/40 = 1/2$. Then, divide Δt_3 by Δt_2 and you get $10/20 = 1/2$. No matter which pair of successive Δt values you divide you will always get $1/2$. One-half is the *rate constant*. It is the same concept as a fixed interest rate on bank and credit accounts.

The rate of change that depends on time only is called *first-order rate* and the kinetics describing such change is *first-order kinetics*. There are also second-order rates, third-order rates, and, particularly in chemical kinetics, the rate orders known as zero-order and three-halves order. We will not trouble ourselves with these types of kinetics. Knowing first-order kinetics will help you understand most of the rates of the processes in the world around you. In chemistry, the rate of a reaction is determined by measuring the amount (mass, number of moles, concentration) of a reactant, R, at any moment of time (just like measuring the temperature every 5 min in the example of making a tea). At the beginning, $t = 0$, there is 100% of the reactant, $R_0 = 1$. As the reaction starts the amount of the reactant goes down and we can express this change by the following relation:

$$R(t) = R_0 \times \exp[-k \times t] \qquad (10\text{-}1)$$

Again, k is the measure of the reaction rate: the higher the rate constant the faster the decrease in the amount of the reactant R. This is similar to money matters: today you have \$100, tomorrow it will be \$50, and on Wednesday it will go down to \$25. The rate constant in this case, as in (10-1), is *negative*: we are monitoring a negative change. In many physical processes it is easier to measure the opposite process – a growth or an increase. The dynamics of heat transfer, as in the example of making a tea, is one of such processes. In chemistry, this would correspond to measuring the change of product, P, in time. At the beginning of a reaction, $t = 0$, there is no product, $P_0 = 0$. At the end, we have all of the product we can get from this reaction, $P(t) = P_{max}$. Like in the case given above, when we were measuring how fast the water in a pot gets hot, a good mathematical relation to express this changes is given as

$$P(t) = P_0 + P_{max} \times [1 - \exp(-k \times t)] \qquad (10\text{-}2)$$

Equation (10-2) is also called an exponential saturation equation.

Problem 10.1 | The joys of sophomore organic lab.

The reactant tertiary butyl chloride, t-BuCl, in aqueous solvent exchanges the Cl^- group with an HO^- to give tertiary butanol, t-BuOH, as the product. The reaction rate is determined by the rate of formation of carbocation intermediate, t-Bu^+ and is first order in t-BuCl. The disappearance of the reactant can be monitored by electrical conductance, the speed at which ions formed in the reaction travel through solution, or, approximately, by using an acid–base indicator. The reaction starts by mixing t-BuCl with solvent to make a 0.020 M concentration at $t = 23°C$. After 45 s the concentration of t-BuCl is 8.13×10^{-3} M and after 2 min 15 s, it is down to 1.31×10^{-3} M. (A) Use these data to calculate the reaction rate, $k[s^{-1}]$, of the hydrolysis of t-BuCl to t-BuOH. (B) How much time [s] would it take for the reactant to drop to 1/2 of the initial concentration?

» Solution – Strategy

Let us see what we have here:

- We know that this is first-order kinetics
- The initial concentration of the reactant, t-BuCl, is given
- The concentration of the reactant is given at $t_1 = 45$ s
- The concentration of the reactant is given at $t_2 = 2$ min 15 s
- The first question is, What is the reaction rate, k?
- Another question is, How much will it take to reduce the initial concentration of t-BuCl to half?

You have everything you need to solve this riddle. Start by writing the formula for the first-order kinetics; since we are monitoring disappearance of the reactant you will use the expression with negative rate constant, $-k$:

$$R = R_0 \times \exp[-k \times t]$$

» Solution A – Calculation

Insert the first set of numbers, for the change after 45 s:

$$8.13 \times 10^{-3} = 0.020 \times \exp[-k \times 45]$$

An efficient way to solve this equation is to move R_0, 0.020 mol, to the denominator on the left side and apply the inverse function, ln, to both sides of the equation; ln(exp) will cancel each other to give you the argument

$$\ln(8.13 \times 10^{-3}/2.00 \times 10^{-2}) = \ln\{\exp[-k \times 45]\} = -k \times 45$$

Multiply the equation through by –1 and you will get the reaction rate:

$$k = -\ln\{8.13 \times 10^{-3}[mol\, L^{-1}]/2.00 \times 10^{-2}[mol\, L^{-1}]\}/45[s] = 0.020[s^{-1}]$$

Note that the unit for k is a reciprocal second. Why? Look back at the original equation, $R = R_0 \times \exp[-k \times t]$. The [argument] of the exp function, that is, everything within the square bracket, *has* to be dimensionless. So if time is given in seconds the first-order rate constant will have to be in reciprocal seconds. What will the unit for k be if time is given in years, [yr]?

Note: You have more data in this riddle than you need. You can take the other set of data, $R = 1.31 \times 10^{-3}$ mol L^{-1}, at 2 min 15 s, $t = 135$ s, and will probably get the same or similar number for the rate constant.

» Solution B – Strategy and Calculation

What do you have to calculate here? Reaction time – the time needed for the initial amount (concentration) of the reactant to drop to one-half. I suggest you first write down the equation for the first-order kinetics

$$R = R_0 \times \exp[-k \times t]$$

Now make a list of the things you know and a list of the things you need to calculate:

1. You know the initial concentration of the reactant, $R_0 = 0.020 \times 10^{-3}$ mol L^{-1}
2. You know the rate constant, $k = 0.020$ s^{-1}
3. You need to calculate, $R(1/2)$, half of the initial concentration
4. You need to calculate $t(1/2)$, the time needed for R_0 to drop to $R(1/2)$.

Let us re-write the above equation, using the $R(1/2)$ and $t(1/2)$ terms, and see if we can simplify some things:

$$R(1/2) = R_0 \times \exp[-k \times t(1/2)]$$

In order to calculate $t(1/2)$ you should do the following: (1) move R_0 to the left side of the equation, (2) take the natural logarithm of the equation, and (3) multiply the equation through by -1. You will then have the following relation:

$$-\ln\{R(1/2)/R_0\} = k \times t(1/2)$$

Now stop for a moment and think how much $R(1/2)$ is. It equals $1/2\ R_0$, right? Insert this and cancel what can be canceled:

$$-\ln\{1/2(R_0)/R_0\} = k \times t(1/2)$$

$$-\ln\{1/2\} = k \times t(1/2) \tag{10-3}$$

Find the ln of $1/2$ and move k to the left side of the equation and you will have for $t(1/2)$

$$t(1/2) = 0.693/k = 0.693/0.020[\text{s}^{-1}] = 34.66[\text{s}] \approx 35[\text{s}]$$

So in 35 s the reactant – no matter how much there was of it in the beginning – will drop to one-half amount. The time needed for this to happen does not depend on the initial concentration. Why? Because the initial concentration, R_0, gets canceled in the equation. Like the rate constant, $t(1/2)$ or $t_{1/2}$ is an important reaction parameter; we call it the reaction *half-life*. Half-life is used in many areas, for example, in enzyme kinetics, radioactive decay, and genetics. Let us try an example in evolutionary genetics.

Problem 10.2	The mitochondrial Eve or – who did you say was your great-great-uncle?

Every once in a while one of the letters in our genetic code, A, C, G, T, gets substituted. This is like when you make a typo in a paper you are writing; you type an A instead of a T, on the third page, seventh sentence and you print the paper. Similar changes, or single nucleotide polymorphisms, SNPs ("snips"), occur in our DNA in an apparently random way and they accumulate over time. Rare as they are, these changes occur more frequently in the mitochondria, mtDNA (in the so-called HR2, hypervariable region 2, in the D-loop of mtDNA), the part of our cells inherited through maternal line [1]. So counting the different SNPs between you and I can be used to trace back in time our "common mother."

It is believed that 10.9 such changes of the total of an average of 16,569 base pairs (bp) in the mtDNA correspond to about 163,000 years, or 163 kyr, on the human evolutionary scale [2]. In other words two human populations which differ on average in 10.9 mtDNA SNPs have separated 163 kyr ago [3]. (You may think of two tribes falling out over some issue and separating 163,000 years before present, BP; it's a little bit like when you are leaving your hometown and going away to a college.)

(A)   Assuming first-order kinetics for the rate of occurrence of random mtDNA SNPs, estimate the number of mtDNA SNPs between today's European population *Homo sapiens* and Neanderthals, *Homo neanderthalensis*, who split away about 465 kyr ago and are believed to have become extinct, Fig. 10.2

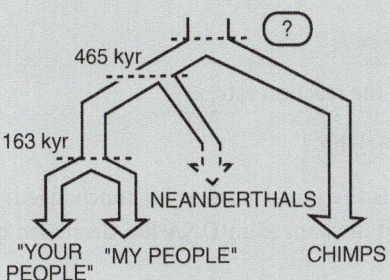

🔹 **Fig. 10.2**   The phylogenetic tree for *Homo sapiens, Homo neanderthalensis* and *Pan troglodytes*

(B)   Given that there is an average difference of 93.4 mtDNA SNPs between *Homo* and that of *Pan troglodytes* (chimpanzee) [2] estimate the time in the past when these two groups split off, as is believed today, from the common primate ancestor, Fig. 10.2 (Now that was a major rift, I would say.)

» Solution A – Strategy

Here is a riddle involving a lot of information – some of which you may not be familiar with – and seemingly quite different than the case of hydrolysis of tertiary butyl chloride in water, which we had before. Yet the common link – first-order kinetics – makes these two problems very similar.

In the case of conversion of *t*-BuCl to *t*-BuOH you were given the initial concentration, R_0, of the *reactant*. After the reaction started you were recording the concentration of the reactant, $R(t)$, and the time t in seconds. You used these data and applied the first-order kinetics math to find the rate constant, k_1, and the half-life of the reaction, $t_{1/2}$.

In the problem with mtDNA SNPs you are given the number of the original nucleotides: 16,569. We will label it bp(0) (i.e., "base pairs at time zero"). You are also given the number of unchanged nucleotides after 163,000 kyr, the time t_1. This is given as $bp(t_1) = 16,569 - 10.9 = 16,558.1$. Notice that 10.9 and 16,558.1 are *average* numbers; of course you cannot change 0.9 part of one nucleotide or 0.1 part of another. When we finish calculation we will round up the result to the nearest integer. What you have to find now is how many of the same initial base pairs, bp(0) = 16,569, are expected to have changed over 465,000 years, $t_2 = 465$ kyr. So you will find the rate constant using the data for 163 kyr and then insert it back and find the number of base pairs at t_2.

» Solution A – Calculation

Write down the equation for the first-order kinetics:

$$bp = bp_0 \times \exp[-k \times t]$$

Solve the equation for the reaction rate, k:

$$k = -(1/t) \times \ln[bp(t)/bp_0]$$

What is bp_0 here? It is the initial number of (unchanged) mtDNA: 16,569. After 163 kyr the number of unchanged mtDNA has decreased by nearly 11 base pairs: $bp(t) = 16,569 - 10.9$.

Insert the numbers given above, punch few keys on your calculator, and you will get for k

$$k = -(1/163,000[\text{yr}]) \times \ln[(16,569 - 10.9)/16,569]$$

$$k = 4.0372 \times 10^{-9} [\text{yr}^{-1}]$$

This is a very small number for a rate constant, wouldn't you say?

Now you will apply the same equation but in the opposite way: you will insert the value of the reaction constant to calculate the number of unchanged base pairs after $t_2 = 465$ kyr, $bp(t_2)$:

$$bp(t_2) = bp_0 \times \exp[-k \times t_2]$$

$$bp(t_2) = 16{,}569 \times \exp\{-4.0372 \times 10^{-9}[\text{yr}^{-1}] \times 4.65 \times 10^5[\text{yr}]\} = 16{,}569$$
$$\times 0.99812$$

$$bp(t_2) = 16{,}538$$

So – what is the answer to the question about the expected number of mtDNA mutations (SNPs) that separate us from Neanderthals? It is the difference between bp_0 and $bp(t_2)$

$$\# SNP_s(\text{Neanderthals}) = bp_0 - bp(t_2) = 16{,}569 - 16{,}538 = 31.07 = 31$$

The 31 SNPs are in the range of the values accepted today though it is a little low; the more common value is 35.3 SNPs [2]. So if your mitochondrial DNA differs from mine by 35 SNPs one of us is likely to be a Neanderthal. Now for the chimps.

» Solution B – Strategy and Calculation

The rate of change is still considered to be first order. The $bp_0 = 16{,}569$, $bp(t_2) = (16{,}569 - 93.4)$, and the rate constant k (4.0372×10^{-9} yr^{-1}) are known. All you have to calculate is the time – let us label it t_3 – when we and the chimps split:

$$bp(t_2) = bp_0 \times \exp[-k \times t]$$

$$t_3 = -(1/k) \times \ln(bp/bp_0)$$

Insert the numbers, punch the keys on your calculator, and you will get

$$t_3 = -(4.0372 \times 10^{-9})^{-1} \times \ln[(16{,}569 - 93.4)/16{,}569]$$
$$= 1.400 \text{ Myr ("Mega year")}$$

So, according to our calculation, the latest that we and chimps were "the same" was about 1.4 million years ago. That is a little upsetting and seemingly incorrect. The number considered to be closer to the human evolutionary models used today could be anytime between 4.38 and 8.37 Myr before present [4]. The humans and chimpanzees are believed to have split off from the common primate ancestor some 6 million years ago. A little safer distance, wouldn't you say?

A comment on mtDNA clock: The method of using mitochondrial DNA for timing matrilineal inheritance (the mother, grandmother, great-grandmother, ... in your family) is a very powerful tool in modern paleontology. The analysis of similarities and differences in nucleotide polymorphisms has brought an unprecedented insight into the evolution of agricultural plants, domestic and wild animals, and humans. Yet the method is rather new and still being tested and challenged by new discoveries of mutation rates. The numbers obtained above, 400–600 kyr for the MRCA (most recent common ancestor) for humans and Neanderthals and 150–200 kyr for the MRCA of the major human groups existing today, though commonly accepted, are being challenged by new findings [3]: there is a strong indication that the very *rate of mutations* is changing (i.e., accelerating). In mathematical terms this would mean that the rate of occurrence of mtDNA SNPs is not first order. In paleoanthropological terms this would mean that human race is not as old as we believe it is today. More interesting discoveries and better insights are likely to be gained in this area in the near future.

Problem 10.3 | The little N-bomb in your heart.

The less well-known plutonium isotope, $^{238}Pu_{94}$, undergoes the following radioactive decay reaction:

$$^{238}Pu_{94} \rightarrow {}^{234}U_{92} + [He]^{+2}$$

This is a highly exothermic nuclear reaction producing 0.54 kW kg^{-1} plutonium and is used in the so-called RTG (radioisotope thermoelectric generator) power units in interplanetary space probes (e.g., Cassini space probe) as well as in some experimental-type pacemakers. Think of the following problem: a newly manufactured $^{238}Pu_{94}$-powered pacemaker is surgically implanted into a patient. Keeping in mind that the Pu-238 half-life is 87.74 yr calculate how much Pu-238 fuel is needed in order to maintain a minimum of 100 µW output for 25 years?

» Solution – Strategy

This is another example of first-order kinetic process. The process here is radioactive decay of Pu-238 isotope. In this process 1 out of every 5.5 billion Pu-238 atoms emits each second an α-particle (He^{+2} nucleus) and turns into U-234:

$$^{238}Pu_{94} \rightarrow {}^{234}U_{92} + {}^{4}He_2^{+2}$$

Decays of radioactive nuclides – of which there are many on Earth – closely follow the first-order kinetics. The equation you need to use is the same as in previous riddles and the question is similar to the question (B) involving mitochondrial mutations: What will the concentration of the reactant be after a certain time?

The trick here seems to be in the fact that the amounts, or concentrations, or numbers of the reactant, R_0 and $R(t)$, are not given directly but are expressed through their thermal energy output per unit mass. So the energy, E, emitted by Pu-238 is proportional to its mass m ($=R$ of the reactant). Let us make a list of things known and things unknown.

We know the following:

- Half-life of Pu-238 is 87.74 years.
- The energy output of the Pu-238 pellet running the pacemaker will have to be equal to or higher than 100 μW.
- The "reaction" time is 25 years, $t = 25$ yr.
- You also know that 1 kg of pure Pu-238 outputs 540 kW energy, $E_0(1 \text{ kg}) = 540$ kW.

We do not know the following:

- The mass of Pu-238 which in 25 years time will have energy output equal to or better than 100 μW; this is our R_0.
- The radioactive decay rate constant; we will probably have to calculate the decay rate to be able to calculate all other quantities.

Let us try simple thinking:

1. Assume you have 1 kg of fresh Pu-238; at this very moment it outputs 540 kW, or $E(\text{now}) = 5.40 \times 10^5$ W of energy (a *lot!*).
2. After 87.74 years this same kilogram of Pu-238 will output how much? One-half of the present energy output, $E(87.74 \text{ yr}) = 5.40 \times 10^5$ W/2 $= 2.7 \times 10^5$ W. But the question is, What will this output be in 25, not in 87.74, years time?
3. First, you have to calculate the rate constant, k. Once you know k you will be able to figure out how much energy will 1 kg of Pu-238 be outputting in 25 years. But note – the mass of Pu-238 does not change significantly; it will be the same 1 kg in 25 years time, so we write $E(1 \text{ kg}, 25 \text{ yr})$.
4. The output energy of the pellet after 25 years time is known; it is 100 μW. What you do not know is the mass of the Pu-238 pellet running the pacemaker. This mass will be very much the same as in 25 years time, only the radioactivity of Pu-238 will decay.
5. Since the energy output is directly proportional to the mass of Pu-238, the *ratio of energies* after 25 years time will be the same as the *ratio of masses* today or in 25 years: $E(25 \text{ yr})/100 \mu\text{W} = m(1 \text{ kg})/m(\text{pellet})$. From this relation you will calculate $m(\text{pellet})$.

To conclude, you need to calculate two parameters: (1) the rate constant k and (2) the radioactivity of 1 kg Pu-238 in 25 years time. Then you will easily find the mass of the Pu-238 pellet to be implanted today.

» Solution – Calculation

Now let us get to work. Write the first-order kinetics equation using the letter E, energy, instead of R, P, or whatever else:

$$E = E_0 \times \exp[-kt]$$

Solve it for k, using the half-life, $t_{1/2} = 87.74$ yr:

$$1/2 \times E_0 = E_0 \times \exp[-k\,t_{1/2}]$$

Cancel E_0 on both sides, apply ln throughout, then multiply by -1 throughout, and solve for k:

$$k = -\ln(1/2)/87.74[\text{yr}] = 7.90 \times 10^{-3}[\text{yr}^{-1}]; \quad \text{the unit is reciprocal year}$$

Using this constant we calculate that the energy output of 1 kg of Pu-238 will drop off to 0.8208 or 82.08% in 25 years. Its mass, of course, will be almost exactly 1 kg. What did we say in statement #5 above? We said that since radioactivity and energy output are directly proportional to mass then the energy of 1 kg Pu-238 after 25 years should be compared with the energy of the Pu-238 pellet. From this ratio we will find out the mass of the pellet:

10

$$E(25 \text{ yr, 1 kg Pu-238})/E(25 \text{ yr, pellet Pu-238}) = \text{mass}(1 \text{ kg Pu-238})/m$$
(pellet Pu-238)

The energy output of the pellet after 25 years is 100 μW or 1.0×10^{-4} W and the energy output of the 1 kg of Pu-238 after 25 years is, as we just calculated, 4.432×10^5 W. When you insert these numbers in the previous equation you will get

$$4.432 \times 10^5 \text{W}/1.0 \times 10^{-4}\,\text{W} = 1 \text{ kg}/m(\text{pellet})$$

This will give the mass of the Pu-238 pellet needed to run a pacemaker:

$$m(\text{pellet}) = 2.256 \times 10^{-10}\,\text{kg} = 2.3 \times 10^{-7}\text{g}$$

A rather small pellet, wouldn't you say? But, given that Pu-238, an artificial element produced using a nuclear reaction, is very expensive and its total yearly use in the world is probably around 40 kg this does not look so small after all.

10.1 Enzyme Kinetics

Enzymes, as you probably know, are proteins that can make chemical reactions happen in a more selective and faster way. At the end of each reaction cycle the enzymes remain unchanged so they act as *catalysts*. Since they occur in the living world we call them biocatalysts. Aside from proteins, ribonucleic acids and their fragments can act as catalysts and are called *ribozymes*, by analogy to enzymes. Enzymes are extracted from living tissues, for example, milk, saliva, liver, muscle; have to be stored under carefully maintained conditions; and, once outside living tissue, lose their activity fast. Isolation and purification of enzymes and assaying their activity have been major operations in biochemical and biomedical laboratories. Today, enzymes are increasingly supplied through use of the techniques of recombinant molecular biology.

Much has been done, said, and written in enzyme kinetics and I will mention only a few things. The enzymes are usually selective; they catalyze only a single reaction or only one type of reaction. While enzymes generally speed up the reactions, in comparison to the same reaction conducted in the laboratory, the enzymes that are not very selective are usually relatively slow. On the other hand, certain highly selective enzymes, like carbonic anhydrase or glutamate mutase, can speed up the reaction conducted under laboratory conditions by a factor of 10^{12}–10^{15}, that is, trillion- to quadrillion-fold. No man-made catalyst matches this efficiency. Thousands of enzymes are known today; they are catalogued into six major categories, in relation to the type of chemical reaction they catalyze. Each enzyme is identified by its enzyme code number, or E.C. number [5].

A reaction between an enzyme, E, and substrate, S, to give a product, P, starts with binding of substrate to enzyme to form a complex, $E{:}S$. This is similar to the interaction of ligand and receptor, $L + R = L{:}R$, that we encountered before. The strength of this complex, expressed by an equilibrium constant, and the rate of conversion of $E{:}S$ into product, expressed by a kinetic constant, are two major parameters used to describe kinetic properties of an enzyme. The mathematical formalism used for enzyme kinetics today has been developed by North American chemists Leonor *Michaelis* and Maud *Menten* and subsequent authors and it is habitually called MM kinetics.

Problem 10.4 | Body builders – check this.

Creatine kinase, CK (or creatine phosphotransferase, E.C. 2.7.3.2) is an important enzyme in our body involved in energy transfer in muscle. Specifically, the enzyme catalyzes phosphate transfer from creatine phosphate, CrP, to Mg^{2+}-coordinated adenosine diphosphate, MgADP, to create adenosine triphosphate, ATP, according to the scheme in Fig. 10.3 .

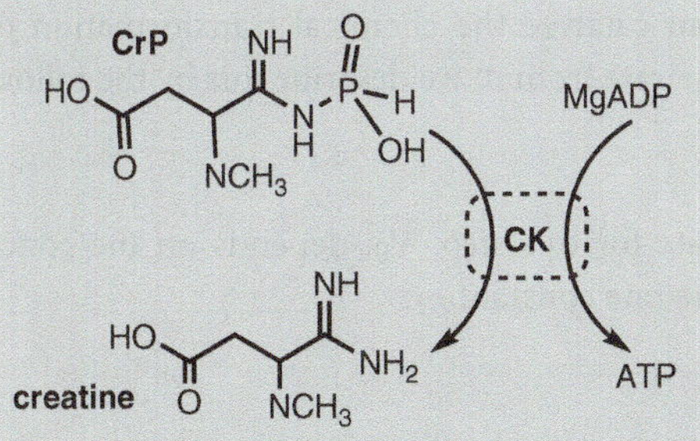

Fig. 10.3 The scheme of conversion of creatine phosphate and MgADP to creatine and ATP, catalyzed by creatine kinase, CK

In an experiment with baby rabbit heart muscle tissue [6], the following kinetic results were obtained: (a) for creatine phosphate, CrP, of concentration 1.8 mM the apparent reaction rate was 1.75 μM min^{-1} and (b) for CrP of concentration 0.35 mM the apparent rate was 0.80 μM min^{-1}. Calculate the Michaelis constant, K_M, and the maximum reaction rate, V_{max}, for this reaction.

» Solution – Strategy

You have substrate, S – which is another word for a reactant – and product, P. If carried out in a laboratory, in a beaker, you can describe this reaction using a simple equation for the reaction rate, RR:

$$\text{RR} = -k[S] \text{ or } \quad \text{RR} = +k[P]$$

As is often the case this reaction may be very slow – taking weeks or years, for example – or it may not even happen in a laboratory. For example, neither substrate nor product is stable enough to stay around during the reaction course. So you need an enzyme, E, and you also have to include it into the rate equation. This is how we go about it.

First, the substrate and the enzyme have to find each other, coordinate in a right way (remember the parables about lock & key or hand & glove?) and bind to make an enzyme–substrate complex, E:S:

$$E + S \rightarrow E{:}S \tag{10-4}$$

The reaction rate, RR, for this interaction is expressed using the reaction constant k_1:

$$\text{RR}_1 = k_1 \times [E{:}S] \tag{10-5}$$

Once formed an $E{:}S$ complex can do three things:

(1) The enzyme can catalyze the chemical transformation of the substrate into product and separate from it; we describe this by the following equation:

$$E{:}S \rightarrow E{+}P \tag{10-6}$$

The reaction rate for this step, V_0, depends on the concentration of the $E{:}S$ complex and the rate constant k_2:

$$V_0 = k_2 \times [E{:}S] \tag{10-7}$$

(2) The $E{:}S$ complex can fall apart back to E and S. This is not very good for the reaction but it happens all the time. You may write for the reaction rate

$$\text{RR}_{-1} = k_{-1} \times [E{:}S]$$

(3) The $E{:}S$ complex does not change either way. The substrate does not turn into product and, more importantly, the enzyme, the catalyst that should catalyze many more reaction cycle molecules before the end of the day, remains locked and inactive. We say that the enzyme is inhibited from further reaction and that S is an inhibitor, I. This happens relatively often, by design or error – you may think of a bad food that just stays in your stomach, doing apparently nothing – and there is a whole little science developed around enzyme inhibitors. Enzyme inhibitors are an important area in the research aimed at finding new medicinal drugs capable of stopping and disabling harmful enzymes, for example, the HIV protease in the human immunodeficiency retrovirus. In terms of enzyme kinetics you may think of it as a dead branch of another reaction of enzyme and (wrong) substrate and write

$$E + I \rightarrow E{:}I$$

Here I stands for inhibitor. The reaction rate of formation of enzyme–inhibitor complex will be

$$RR(I) = k(I) \times [E{:}I]$$

Now – back to our problem. In the riddle about creatine kinase, CK, you need to consider only the first two possibilities, the formation of the enzyme–substrate complex and the conversion of substrate to product. You write the reaction equation:

$$E + S \leftrightarrow ES \rightarrow P + E$$

The rates will include k_1, k_{-1}, k_2 and $[S]$, $[E{:}S]$ and $[P]$

$$k_1([E]_{tot} - [E{:}S])[S] = k_{-1}[E{:}S] + k_2[ES] \tag{10-8}$$

In the equation above E_{tot} is the total concentration of enzyme (total amount, if we are talking about the chemicals dissolved in the same volume of solvent). Like in the case of ligand–receptor, under the chapter on chemical equilibrium, we define it as a sum of free enzyme and enzyme bound in complex: $[E]_{tot} = [E]_{free} + [E{:}S]$. Of course you want to have as little $[E]_{free}$ as possible. There are also a number of assumptions and simplifications used in developing the relations that help us process the enzyme kinetics data.

≫ Assumptions

- We can usually – but *not* always – make the concentration of the substrate so much higher than the enzyme's that the kinetics of the formation of the $E{:}S$

complex, which depends on two reactants and is *second order*, RR = $k[E][S]$, becomes now dependent on one reactant only, RR $\approx k[E]$. We call this experiment a *pseudo-first-order* kinetics. The simplification from higher order to first order using the pseudo-first-order kinetics is used extensively in chemistry and biochemistry. (This is similar to the water solvent in acid–base reactions; there is so much of it that we assume its concentration to be constant.)

- We assume the *steady-state approximation*, that is, once the reaction flow $E + S \leftrightarrow ES \rightarrow P + E$ establishes, the ratio of k_1, k_{-1}, and k_2 becomes constant. As a consequence, [E:S], the concentration of the enzyme–substrate complex, is assumed to be constant throughout the reaction.

You may reshuffle the equation given above into the form

$$k_1[E]_{tot}[S] - k_1[E:S][S] = (k_{-1} + k_2)[E:S] \tag{10-9}$$

Like in most other physical chemical topics this is not about math but about using math to describe what is known and what can be experimentally accomplished. In an enzyme kinetics experiment, measuring the concentration of the enzyme–substrate complex, [E:S], is usually difficult if not impossible so you try to re-write the whole expression in terms of the quantities which are known, [S] and $[E]_{tot}$, for example, or which can be measured, the rate of reaction at maximum [S], V_{max}, for example. Keeping this in mind you can reshuffle (10-9) to obtain an expression for [E:S]:

$$[E:S] = [E]_{tot}[S]/([S] + (k_{-1} + k_2)/k_1) \tag{10-10}$$

Now an important substitution can be made after you define K_M:

$$K_M = (k_{-1} + k_2)/k_1 \tag{10-11}$$

K_M is the so-called Michaelis constant; an equilibrium *thermodynamic* constant expressed through *kinetic* constants, k_1, k_{-1}, and k_2. The previous equation now reads

$$[E:S] = [E]_{tot}[S]/([S] + K_M) \tag{10-12}$$

Back, at the very beginning of this long derivation we defined initial reaction speed as the rate of decomposition of the E:S complex into enzyme and product: $V_0 = k_2 \times [E:S]$. You can now re-write this equation by inserting the expression for [E:S] from the previous equation

$$V_0 = k_2 \times [E]_{tot}[S]/([S] + K_M) \tag{10-13}$$

You need one more thing: When the concentration of substrate is so high that almost no free enzyme is left in the solution, that is, when $[E]_{tot} \approx [E:S]$, the initial rate, $V_0 = k_2 \times [E:S]$, reaches the maximum reaction rate, V_{max}. This will give you the enzyme kinetics equation known as the Michaelis–Menten or the MM equation:

$$V_0 = V_{max}[S]/([S] + K_M) \tag{10-14}$$

» Solution – Calculation

If you read the riddle again you will find two substrate concentrations, $[S]_1 = 1.80$ mM and $[S]_2 = 0.35$ mM, and two *apparent* reaction rates, $v_1 = 1.75\,\mu$M min^{-1} and $v_2 = 0.80\,\mu$M min^{-1}. The *apparent rate* is a cautious word for the observed, measured reaction rate. I suggest you write the MM equations with these data and then see what you know and what has to be calculated. You will have

$$1.75 \times 10^{-6} = V_{max} \times 1.80 \times 10^{-3}/(1.80 \times 10^{-3} + K_M) \tag{A}$$

$$0.80 \times 10^{-6} = V_{max} \times 0.35 \times 10^{-3}/(0.35 \times 10^{-3} + K_M) \tag{B}$$

So you have two equations with two unknowns, V_{max} and K_M. An old trick is to make these two equations as similar as possible, subtract them, and remove one unknown parameter. First, multiply what has to be multiplied and re-arrange the equations:

$$1.75 \times 10^{-6} \times 1.80 \times 10^{-3} + 1.75 \times 10^{-6} \times K_M = V_{max} \times 1.80 \times 10^{-3} \tag{A}'$$

$$0.08 \times 10^{-6} \times 0.35 \times 10^{-6} + 0.80 \times 10^{-6} \times K_M = V_{max} \times 0.35 \times 10^{-3} \tag{B}'$$

Now you can multiply (B)$'$ by 1.80×10^{-3} and divide it by 0.35×10^{-3} and you will get

$$1.44 \times 10^{-9} + 4.11 \times 10^{-6} \times K_M = V_{max} \times 1.80 \times 10^{-3} \tag{B}''$$

Now you can subtract (B)$''$ from (A)$'$ to eliminate $V_{max} \times 2.1$:

$$(A)' - (B)'' = (3.15 \times 10^{-9} - 1.44 \times 10^{-9}) + (1.76 \times 10^{-6} - 4.11 \times 10^{-6}) \times K_M = 0$$

From here you get the Michaelis constant, K_M:

$$K_M = 1.71 \times 10^{-9}/2.35 \times 10^{-6} = 7.28 \times 10^{-4} \text{ M}$$

Now, when you insert the value for K_M into either MM equation you will get for V_{max}

$$V_{max} = (1.75 \times 10^{-6} \times 1.80 \times 10^{-3} + 1.75 \times 10^{-6} \times 7.28 \times 10^{-4})/1.80 \times 10^{-3} = 2.46 \times 10^{-6} \text{ mol min}^{-1}$$

A comment on K_M: Michaelis constant, an often cited parameter, is indicative of both thermodynamic and kinetic properties of an enzyme-catalyzed reaction; itself, however, K_M is a measure of neither. Given that it contains the rate constant for the enzyme-substrate association in the denominator, $K_M = (k_{-1} + k_2)/k_1$, (where rate $= k_1[E][S] \approx k_1[E]$ for $[S] >> [E]$), Michaelis constant is inversely proportional to the affinity of an enzyme for substrate; the higher the rate of this reaction, the lower the K_M.

10.2 Reaction Barriers

We have seen before in the section on thermodynamics that the standard Gibbs energy, $\Delta G^{\ominus}$, tells us whether a certain reaction is spontaneous, $\Delta G^{\ominus} < 0$, or whether it has to be driven (by temperature, light, or other agency), $\Delta G^{\ominus} > 0$.

However, the $\Delta G^{\ominus}$ does not provide *any* information about how fast a reaction will happen. The rate of reaction is related to the activation energy (or reaction energy barrier). You may think of it as a hurdle that reactants have to jump over. We express a relation between a reaction rate constant, k, and a reaction activation energy, E_a, using the so-called Arrhenius equation, after the Swedish physicist Svante Arrhenius:

$$k = A_0 \times \exp[-E_a/RT] \tag{10-15}$$

E_a, more correctly written as ΔG_a, or $\Delta G^{\ddagger}$, the Gibbs activation energy, resolved into the enthalpic and entropic contributions, $G^{\ddagger} = H^{\ddagger} - T \times S^{\ddagger}$, is the reaction energy barrier. Here, the double-dagger symbol, $\ddagger$, refers to the *transition state*: a moment in a reaction defined by an intermediate state. This is a state when the reactant molecules start falling apart but the product molecules are not yet fully formed. This state could be a real chemical species or just a theoretical model. Again, it is very important to understand what the two Gibbs energies are telling us, so we show it in detail in the following graphs, Fig. 10.4 and Fig. 10.5:

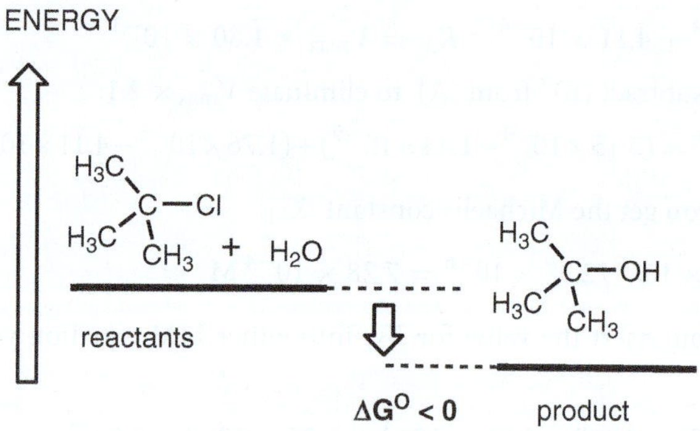

Fig. 10.4 Change in $\Delta G^{\ominus}$ for a spontaneous reaction

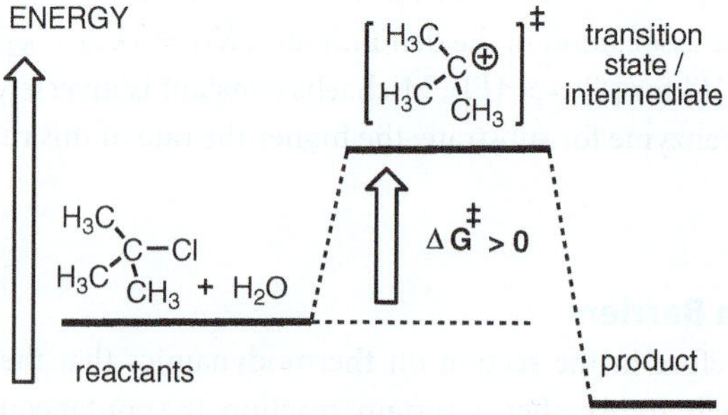

Fig. 10.5 Activation energy barrier is always positive

A_0 or *Arrhenius factor* is a fitting coefficient; it is usually unknown and difficult to determine or even calculate (using theoretical estimates of a likelihood of two molecules engaging in a reaction). It is better if we can eliminate it from a problem. The ΔG^Θ determines the likelihood of the reaction: $\Delta G^\Theta < 0$ for spontaneous reaction, Fig. 10.4, and $\Delta G^\Theta > 0$ for a driven reaction (not shown). The activation energy is always positive, $\Delta G^\ddagger > 0$, and determines the *rate* of the reaction, Fig. 10.5; ΔG and $\Delta G^\ddagger$ are unrelated.

Without going into more detail at this point, I will continue to call the activation Gibbs energy just *activation energy* and use the common symbol E_a. In the majority of cases the Arrhenius equations are used to compare two reaction rates at two different temperatures, or other reaction conditions:

$$k_1 = A_0 \times \exp[-E_a/RT_1]$$

$$k_2 = A_0 \times \exp[-E_a/RT_2]$$

In the next step you divide one equation by another:

$$k_1/k_2 = A_0/A_0 \times \exp[-E_a/R(T_1 - T_2)] \tag{10-16}$$

After canceling A_0, the "Arrhenius factor," you solve the equation for E_a. In this way you can calculate, or estimate, reaction activation energy which is an important but difficult to determine parameter:

$$E_a = -R(T_1 - T_2) \times \ln k_1/k_2 \tag{10-17}$$

Another, often-cited, version of the Arrhenius equation is its logarithmic form:

$$\ln k = \ln A_0 - E_a \times (1/RT) \tag{10-18}$$

The semblance to a linear equation, $y = b - ax$, is used to graphically determine the slope, E_a, and even the – highly approximate – intercept, $\ln A_0$, from a graph constructed using several reaction rate constants determined at sufficiently different temperatures. This procedure has been shown to be of limited practical value for it is usually difficult to determine reaction rate constants over a sufficiently wide temperature range, particularly true for biological reactions which take place only within a narrow temperature span.

| Problem 10.5 | A tribute to Svante (sweet or sour). |

Xanthine dehydrogenase, XDH [E.C. 1.1.1.204], and the closely related isoenzyme xanthine oxidoreductase, XOR [E.C. 1.1.3.22], are molybdopterin oxidoreductive enzymes. The enzymes catalyze oxidation of a C-substrate atom by two electrons. In human body the enzymes convert the nucleic acid metabolite hypoxanthine into xanthine and xanthine into uric acid, Fig. 10.6 The reductive rate constant of XDH,

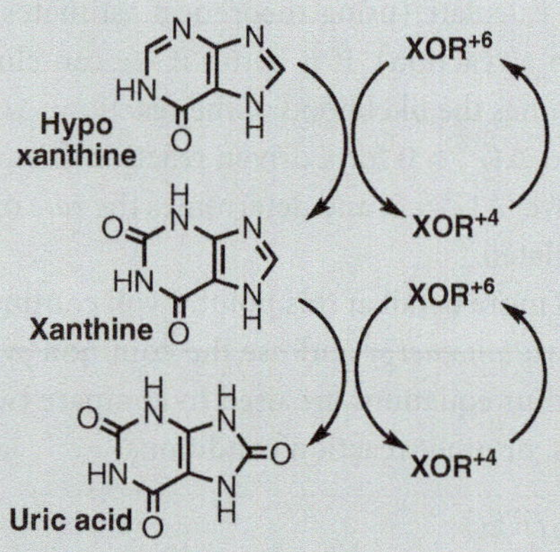

■ **Fig. 10.6** The consecutive oxidoreductive reaction steps, catalyzed by xanthine oxidore-ductase in our body

k_{red}, for xanthine substrate was found to be 67 [s^{-1}] [7]. On the other hand, from a simple chemical model of this reaction step, we obtain an activation energy of 76 kcal mol^{-1} [8, 9]. According to a simplest model, we say that the 76 kcal mol^{-1} energy barrier is what determines the $k_{red} = 67$ [s^{-1}]. In an improved model that includes acid–base properties of the enzyme active site, the activation energy for the same reaction step is lowered by 10 kcal mol^{-1}. Assume pseudo-first-order kinetics and use the activation energies and the rate constant data to calculate the *change* in the reaction rate induced by the 10 kcal mol^{-1} activation energy drop.

≫ **Solution – Strategy and Calculation**

This is a typical problem requiring the Arrhenius equation: two reaction rates at two reaction conditions. However, unlike in the majority of cases, the temperature here remains the same for both reactions but the activation energy, E_a, changes. So you can write

$$k_1 = A_0 \times \exp\left[-E_a(1)/RT\right]$$

$$k_2 = A_0 \times \exp\left[-E_a(2)/RT\right]$$

You now have two algebraic linear equations with two unknowns, k_2 and A_0; you are interested in k_2 only. If you divide one equation by another the A_0 terms will cancel and you will be left with only the unknown, the rate constant, k_2, at the lower activation energy. In fact you need only the *ratio* of the rate constants, k_2/k_1:

$$k_2/k_1 = A_0 \times \exp\left[-E_a(2)/RT\right]/A_0 \times \exp\left[-E_a(1)/RT\right]$$

$$k_2/k_1 = \exp[-[-\text{Ea}(2)/RT]/A_0 \times \exp[-E_a(2)/RT]$$
$$= \exp[-(E_a(2) + E_a(1))/RT]$$

Note that $E_a(1)$, when moved up, changes the sign, in accordance with the rule for exponents: $10^{-1/x} = 10^{+x}$. You do not really have to know k_2, the question asks you to find the *change* of the k_2 and k_1 *ratio*. So you write

$$k_2/k_1 = \exp[-(E_a(2) + Ea(1))]/RT$$

Insert the numbers for R, T, and the two activation energies, $E_a(1) = 76$ kcal mol$^-$ and $E_a(2) = 66$ kcal mol^{-1}.

Conversion of units : You have to be careful now as the activation energies are given in kcal mol^{-1}, that is, in units of 1,000 calories per mole while the gas constant, R, is given in J mol^{-1}. You will have to convert both activation energies to kJ mol^{-1} by multiplying them by 4.187 (4.184 in older literature) and then multiplying them by 1,000 to bring them to the same level as the units for R, the gas constant in J mol^{-1}:

$$k_2/k_1 = \exp\{(-\{(-66[\text{kcal mol}^{-1}] + 76[\text{kcal mol}^{-1}])$$
$$\times 4.187[\text{kJmol}^{-1}\text{kcal mol}^{-1}]$$
$$\times 1,000/8.314\,\text{J}[\text{mol}^{-1}\text{K}^{-1}] \times 298.15[\text{K}]\}$$
$$k_2/k_1 = \exp[16.89] = 21,662,810 = 2.17 \times 10^7$$

So the 10 kcal mol^{-1} drop in activation energy for this reaction is predicted to result in an almost 22,000,000-fold increase in the rate constant – a large reaction rate acceleration. While this number may not be completely correct it does provide an insight into the means by which catalysts, enzymes in particular, can speed up chemical reactions.

References

mtDNA Clock
1. Schatz G, Haslbrunner E, Tuppy H (1964) Deoxyribonucleic acid associated with yeast mito-chondria. Biochem Biophys Res Commun 15:127–132
2. Krings M, Geisert H, Schmitz RW, Krainitzki H, Pääbo S (1999) DNA sequence of the mito-chondrial hypervariable region II from the Neanderthal type specimen. PNAS 96:5561–5565
3. Ho SYW, Phillips MJ, Cooper A, Drummond AJ (2005) Time dependency of molecular rate estimates and systemic overestimation of recent divergence times. Mol Biol Evol 22:1561–1568
4. Kumar S, Filipski A, Swarna V, Walker A, Hedges SB (2005) Placing confidence limits on the molecular age of the human–chimpanzee divergence. PNAS 102:18842–18847
5. BRENDA The Comprehensive enzyme information system: http://www.brenda-enzymes.org/, Accessed July 31, 2009

Creatine Kinase

6. Perry SB, McAuliffe J, Balschi JA, Hickey PR, Ingwall JS (1988) Velocity of the creatine kinase reaction in the Neonatal rabbit heart: role of mitochondrial creatine kinase. Biochemistry 27:2165–2172

Xanthine Oxidase

7. Leimkühler S, Stockert AL, Igarashi K, Nishino T, Hille R (2004) The role of active site glutamate residues in catalysis of Rhodobacter capsulatus xanthine dehydrogenase. J Biol Chem 279:40437–40444
8. Ilich P, Hille R (1999) Mechanism of formamide hydroxylation catalyzed by a molybdenum-dithiolate complex: A model for xanthine oxidase reactivity. J Phys Chem B 103:5406–5412
9. Ilich P, Hille R (2002) Oxo, thioxo, and telluroxo Mo-enedithiolate models for xanthine oxidase: Understanding the basis of enzyme reactivity. J Am Chem Soc 124:6796–6797

10

Table V

Summary of kinetics – what have we learned in this section?

Review the material we have covered, write down the new words, and describe them. Then re-write these descriptions using symbols and numbers.

Words and Phrases	Symbols, Formulas and Numbers
Substrate, product	$[S]$, $[P]$
Rate of reaction, RR	$RR = -k\,[\text{reactants}] = +k\,[\text{products}]$
First-order kinetics	$RR = -k_1\,[S]; [S] = [S_0] \times \exp[-k_1\,t]$
Reaction half-life	$k_1\,t(1/2) = \ln 2$
Second-order kinetics	$RR = -k_2\,[S_1][S_2]$
Enzyme kinetics: enzyme, E, substrate, S, enzyme–substrate complex, $E{:}S$, product, P	$S + E \rightarrow E{:}S \rightarrow P + E$
Michaelis–Menten equation	$V_0 = k_2 \times [E]_{\text{tot}}\,[S]/([S] + K_M)$
Michealis constant	$K_M = (k_{-1} + k_2)/k_1$
Activation energy, Arrhenius equation	$k = A_0 \times \exp[-E_a/RT]$
Two rate constants at two temperatures	$K_1/k_2 = \exp[-E_a/R \times (T_1{}^{-1} + T_2{}^{-1})]$

Part VI

Structure of Matter: Molecular Spectroscopy

11 **The Structure of Matter** 155

12 **Interaction of Light and Matter** 171

11 The Structure of Matter

The physical chemistry we have presented so far is concerned with objects and actions comparable to the human scale: meter, kilogram, second. Physical chemistry is fairly successful in describing this, *macroscopic,* world. To understand the properties of materials at *atomic scale* you have to reduce the size of everyday objects by billionfold or more, to nanometer (10^{-9} m), piconewton (10^{-12} N), or attosecond (10^{-18} s). Physicists have taught us how to look at this world and we have come a long way toward understanding matter and energy at atomic scale. Then there is the middle world ("Middle Kingdom"), where actions are neither atomic nor macroscopic but *mesoscopic* – from *mesos,* Old Greek for middle – and where objects measure tens or hundreds of nanometers; therefore nanoscience and *nanotechnology.* This is a matter of intensive current study.

Imagining the size or mass of an atom or an electron, or understanding the way they spin or traverse from one point to the next is humanly impossible. However, learning the ways and rules of atoms and molecules is not a pure academic exercise. You need to understand what is a blue color, how does microwave oven heats up your coffee, why is sunset red (and what is "red"?), or what your cell phone might be doing to your brains. You need to understand why household bleach removes stains or how you can help someone poisoned by carbon monoxide stay alive. You need to know the little molecules and atoms in order to understand the big world around you.

11.1 Simple Quantum Mechanics

Macroscopic world – the world we see and feel – is made of solid, liquid, or gaseous mater. Gravity, pressure, force and friction are important in this world. This is also known as deterministic world: all motions of the bodies around us are completely predictable. (Or else there would not be a game of pool.)

Gravity, friction, pressure, or forces in general are not important in the atomic world. The particles in the atomic world should be viewed as packets of action; they can have mass, like proton and electron do, or they can be pure, massless energy, like a photon, a particle of light. As such the atomic particles are also indeterminate;

P.-P. Ilich, *Selected Problems in Physical Chemistry,*
DOI 10.1007/978-3-642-04327-7_11, © Springer-Verlag Berlin Heidelberg 2010

you can tell only what is probable to happen but not what is going to happen. You can know the position, energy, and other properties of an atomic particle only after you engage it in action, after you take a *measurement* of a particle. The mathematics suited for atomic particles is based on observation of waves and description of motion and propagation of waves, therefore the expression wave mechanics of atomic particles.

We cannot afford to engage in a study of even basic principles of microscopic wave mechanics here and now but a few simple ideas and formulas will take us a long way. Probably the most important concept in wave mechanics is energy. Like the energy in our macroscopic world, the energy in atomic world can be described as the energy of motion and the energy of position, that is, kinetic energy and potential energy, respectively:

$$E = E_{kin} + E_{pot}$$

Potential energy in our world is most often caused by gravity; it can also be due to resistance of a spring. The potential energy that keeps together atoms in a molecule is coming from the so-called *Coulombic* interaction: the attraction or repulsion of electrically charged atomic nuclei and electrons.

$$E_{pot} = const. \times q_1 \times q_2 / r_{12}^2$$

Here the symbols have the same meaning as in electricity, q_1 and q_2 are charges, $r_{1,2}$ is the distance between the two charged particles, and const. is a constant placed here to account for different units. Nothing really new there. The term for kinetic energy is more interesting. Like in the macroscopic world kinetic energy is expressed by a moment and a mass:

$$E_{kin} = p^2/2m \tag{11-1}$$

In macroscopic, *classical*, or Newtonian mechanics, $p = m \times v$, where $v =$ velocity. In wave mechanics, however, p is given as

$$p = h/\lambda \tag{11-2}$$

The substitution, $p = h/\lambda$, suggested by French mathematician Louis de Broglie, gives kinetic energy a novel form:

$$E_{kin} = h^2/2m\,\lambda^2 \tag{11-3}$$

This needs an explanation: λ is a wavelength. In quantum mechanics every particle is described as a wave traveling through space. You and I have our wavelengths, only they are very, very short ($\sim 10^{-35}$ m) and probably a negligible part of our daily activities; h is the minimum amount of action – a *quantum* of action. It is indeed small and its unit is joule times second, $h = 6.62606896 \times 10^{-34}$ J s [1, 2]. It is also an important quantity, used in almost every expression involving atoms, electrons, and photons. In honor of the Austrian physicist Max Planck, one of the creators of quantum mechanics, h is called Planck's constant.

Make a note: *Planck's constant.*

The energy of a particle wave – like a wave of light – is expressed by the Planck–de Broglie's relation:

$$E = h\nu \quad \text{or} \quad E = hc/\lambda \tag{11-4}$$

Here h is Planck's constant and ν is the frequency of light, i.e., the number of times the light wave reaches a crest in a unit time. In music, this is the drum beat. For light, ν can be 10^{14} s^{-1} for red light to 10^{18} s^{-1} for x-ray "light." For comparison, you and I can type one word every two seconds (on a good day), so the frequency of our typing is 2 s^{-1}. You see – the properties in the world of atoms and molecules are either extremely small, like the sizes for the atoms and electrons, or extremely large, like frequencies. In (11-4) c is the speed of light, c(vacuum) $= 299,792,458$ m s^{-1}[1, 2]. It is interesting to equate the two energy expressions:

$$E = h\nu \quad \text{and} \quad E = h^2/2m\lambda^2$$

$$h\nu = h^2/2m\lambda^2$$

Or, after dividing both sides with h, we get

$$\nu = h/2m\lambda^2 \tag{11-5}$$

This equation is fundamental for wave mechanics of atomic particles: it shows that a motion of a microscopic particle depends on frequency (reciprocal time) to first order and on wavelength (reciprocal length) to second order. Written in another form, using first-order time derivative, $\partial/\partial t$, and second-order distance derivative, $\partial^2/\partial x^2$, this expression is known as Schrödinger equation, named after the Austrian physicist Erwin Schrödinger. It turns out that, in addition to energy, other properties, like the path of an electron around the nucleus and the magnitude of its moment of rotation, can be expressed in small, indivisible units, *quanta*. So the whole branch of physics and, later, chemistry, biochemistry, and other areas adopted the word quantum: *quantum mechanics, quantum chemistry,* even quantum pharmacology. Today in the era of extensive numerical modeling of natural phenomena and intense networking and information exchange the terms progressively more used are *quantum computing* and *quantum cryptography*. Quantum mechanics is considered the greatest intellectual achievement of human mind [3–7]. Here is a question regarding a simple quantum mechanical model.

Problem 11.1	"…as birds that entering by the chimney, and finding themselves enclosed in a chamber," (Leviathan, IV, Thomas Hobbes, 1651)

Three electrons, $m_e = 9.11 \times 10^{-31}$ kg, occupy the lowest levels of a square potential well of width L, $L = 250$ pm. Calculate their total kinetic energy, E_n.

» Solution – Strategy

Quantum mechanics may sound a little scary – after all, Albert Einstein failed to understand it – but it is a well-designed theory that has so far proven to be correct. The fact is, many quantum mechanical concepts – and models – are rather simple. Such is the model of *square potential well* mentioned in the question. It is a well, an imaginary well, surrounded by walls of high potential energy, V_0, and it contains levels – quantum levels – starting from bottom and going up. Some of these levels could be occupied by electrons, atoms, or other atomic particles. The higher the level a particle occupies the higher its kinetic energy. You may think of these levels as rungs in a ladder; a very tall ladder. In fact the kinetic energy of a particle inside the well, no matter how high a level it occupies, is always smaller than the potential energy of the wall of the well, $V_0 > E$. A particle cannot escape the well. For this reason such well is considered to have infinitely high potential energy walls; also all of its walls are considered to be equally high. Accordingly, it is called a *symmetric infinite potential well*. It is also known as square well, to set it apart from wells of other shapes and dimensions. Another name for this model, used more often by chemists, and including both the well and the particles inside it is *particle in a box*. The well in the question is a simple one: it has left and right sides, both stretching infinitely high, and a number of rungs, that is, quantum levels. The distance between the two sides of this one-dimensional well is L, given as 250 pm. This is a small well – we are talking atoms and electrons here. The numbering of the quantum levels in the well starts with one: $n = 1$, then $n = 2$, then $n = 3$, and so on. The kinetic energy of a quantum object – a microscopic particle that can be described by quantum mechanical laws and equations – inside the well depends on its mass, m, the width of the well, L, and the quantum number, n:

$$E_n = n^2 \times h^2 / 8\,\mathrm{m} \times L^2 \tag{11-6}$$

Here h is of course the Planck's constant. Look at the formula for energy and you will see what we had said before: the higher the level (the rung in the well) the higher the energy. Since the energy depends on the *square* of the quantum number it increases much faster than the number of the rung which it occupies, that is, when the quantum levels increase like 1, 2, 3, 4, . . ., the energy levels will increase like 1, 4, 9, 16, ... The energy of the quantum particle inside a well depends also on how wide the well is; the energy is inversely proportional to the square of the width of the well. So the wider the well the lower the energy levels. You may think of guitar strings: the pitch of the highly strung string will be higher than of a looser string. This is pretty much all there is to the quantum potential well, so let us now churn out some numbers.

» Solution – Calculation

The question here is this: What is the *total energy* of the three electrons in a quantum well of given size? Let us first draw a little scheme, a scheme of a symmetric infinite square potential well (Fig. 11.1(A)), then scheme of a well with one electron at the lowest quantum level (Fig. 11.1(B)), with two electrons (Fig. 11.2(C)), and, finally, with three electrons in the lowest quantum levels (Fig. 11.2(D)):

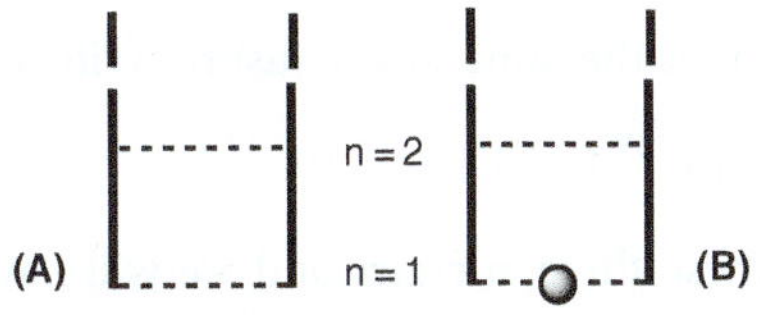

Fig. 11.1 A square well (A) with an electron (B)

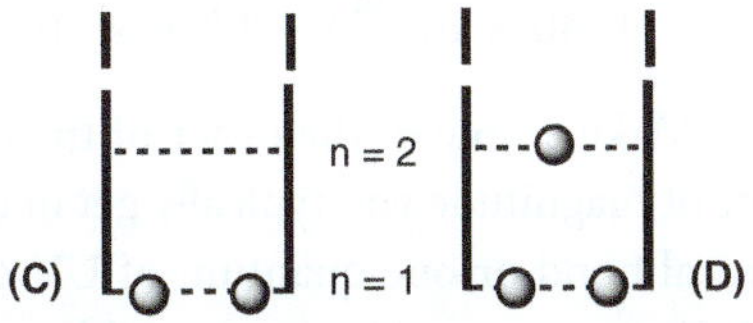

Fig. 11.2 The well with two (C) and three (D) electrons occupying the lowest levels

Now the "infinite" well, (Fig. 11.1(A)), does not really look infinite; you have to use your imagination and think how the walls of the well, broken into shorter and shorter lengths, mean precisely this: the height of the wall is infinite. You place the first electron on the first rung so it gets assigned the quantum number 1, $n = 1$. Let us calculate its energy using the formula given above:

$$E_n = n^2 \times h^2 / 8\,m \times L^2$$

When you insert the numbers for n, h, m, and L you will have the energy:

$$E_n(1) = 1^2 \times (6.626 \times 10^{-34}\,[\text{Js}])^2 / 8 \times 9.109 \times 10^{-31}\,[\text{kg}] \times (250.0 \times 10^{-12}\,[\text{m}])^2$$

$$E_n(1) = 9.640 \times 10^{-19}\,\text{J}$$

The little picture (C) tells us that the second electron is occupying the lowest quantum level, $n = 1$, and you should calculate its energy the same way or – rather, just copy it:

$$E_n(2) = 9.640 \times 10^{-19}\,\text{J}$$

And the third electron, the same? Well, not. A rule about electrons (and protons, neutrons, and many other elementary particles) says that no more than *two* can be placed in the same quantum level. The rule comes from the so-called relativistic quantum mechanics and quantum statistics and is known as *Pauli exclusion principle*, after the German physicist Wolfgang Pauli. As the scheme (D) shows the third electron takes the next rung, $n = 2$, and its energy is given as

$$E_n(3) = 2^2 \times (6.626 \times 10^{-34} \text{ [J s]})^2 / 8 \times 9.109 \times 10^{-31} \text{ [kg]} \times (250 \times 10^{-12} \text{ [m]})^2$$

The rest of the expression is the same so you just re-write it as

$$E_n(3) = 4 \times 9.640 \times 10^{-19} \text{ J} = 3.856 \times 10^{-18} \text{ J}$$

Et – voilà, let us add these three energies and we will have the answer: the total energy of three electrons, occupying the lowest rungs of a square potential well of the width $L = 250$ pm.

$$E_{tot} = E_n(1) + E_n(2) + E_n(3)$$

$$E_{tot} = 9.640 \times 10^{-19} \text{ J} + 9.640 \times 10^{-19} \text{ J} + 3.856 \times 10^{-18} \text{ J} = 5.784 \times 10^{-18} \text{ J}$$

A comment on the result: Make a note of the order of the magnitude of the result: 10^{-18} J. This is the order of magnitude you typically get in calculations of this type; the energy of one chemical bond or one quantum of UV or visible light is in this range: 10^{-18} to 10^{-19} J. If you are working on a problem involving UV/Vis photons and electrons and you are getting numbers larger than 10^{-16} J or smaller than 10^{-21} J for a particle of light or a particle of matter, you are likely doing something wrong. Now – if you are thinking of chemical energies, as given in the textbooks and tables, which are in the 400–1600 kJ range, then you will have to multiply the 10^{-18} J result by the Avogadro number, $N_A = 6.022 \times 10^{23}$ mol^{-1}. This is an important and pesky little thing in physical chemical problems of this type and I will explain it in more detail in one of the following problems.

Make another note: *Energy per particle* and *energy per mole*.

Problem 11.2 | And then there was light.

On the back of your eyebulb, in the cell layer called retina, there are large numbers of small cellular bodies known as rods and cones. They contain the organic chemical *retinal*, housed in a protein called *opsin*; together, retinal and opsin make *rhodopsin* – the human eye light detector. Under the light of a certain wavelength retinal, a strongly conjugated compound, undergoes a change from *cis*- to *trans*-isomer and dissociates from opsin. This act triggers a cascade of chemical processes that, about 1/1,000 second later, send a message to the brain saying: "There is light" [8–10]. Look at the retinal formula, Fig. 11.3, identify and count the conjugated bonds and then count the conjugated carbon, nitrogen, and oxygen atoms. Assuming, first, that the sum of all bond lengths equals the width of a potential well and, second, that there is one delocalized π-electron per conjugated C, N, or O atom, equate the retinal molecule to an *infinite potential well model*.

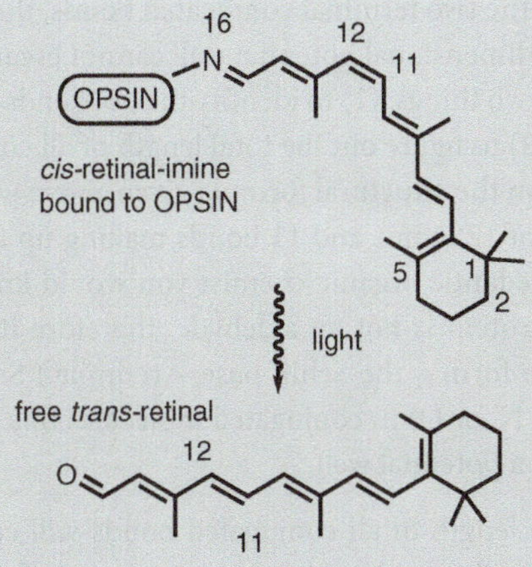

Fig. 11.3 Retinal – the light-detecting molecule in our eyes and its transformation that starts the process of vision

(A) Calculate the total energy of the $\pi(p)$ electrons in retinal. (B) Assume that a photon of light excites one $\pi(p)$ electron from HOMO to LUMO; calculate the HOMO–LUMO energy gap and the wavelength of the exciting light. Compare this result with the green-yellow light of 578 nm to which human retinal is highly sensitive [11].

Solution A – Strategy

Let us do the first part, (A), first. Here you have a one-dimensional potential well – the same type of problem and same math we have had before. There are also few twists, tricks, and new things but we will address them in due order. The path to solving this problem, by using the model of a potential well, is based on the following assumption:

Assumption #1

- The length (width) of the potential well, L, equals *the total length of the conjugated part* of retinal molecule.

This is actually a fairly solid assumption, backed by ample experimental evidence. Let us recall that a conjugated molecule is made of a string of single and double bonds; we say alternating single and double bonds. The $\pi(p)$ electrons in the conjugated part of the molecule are not tied to one atom but move freely from the leftmost to the rightmost atom delimiting the conjugated molecule; we say the $\pi(p)$

electrons are *delocalized* through the length of the molecule. Also, the $\pi(p)$ electron cannot escape beyond the two terminal conjugated bonds, the "bookends," just like the electrons in a one-dimensional potential well cannot escape the well.

Now we have to do two things: (1) to identify the bookends of the conjugated part of the molecules and (2) to figure out the total length of all conjugated bonds.

The end atoms: From the structural formula given above we see these are C5 and N16: there are a total of 12 atoms and 11 bonds making up the conjugated part of retinal. (If you are a pedantic organic chemist you would know that the molecule conjugated with the protein is not an aldehyde, therefore it should not be called retin*al* but is an imino form – the Schiff base – retin*ine*.) Since there is one $\pi(p)$ electron per center (C, N, or O) in conjugated molecules this means we have a total of 12 $\pi(p)$ electrons in a potential well.

The bond lengths: The length of all conjugated bonds will equal the width of our potential well model. In the structural formula above we find six double and five single bonds. These bonds, being a part of conjugated part of the molecule are neither single nor double. We can go back to organic chemistry textbooks or similar databases and search for the lengths of conjugated bonds; alternatively we can use the following assumption.

» **Assumption #2**
- In a conjugated molecule, we assume that the length of the double and the single bonds are sufficiently well approximated by a single value – the length of the C, C bond in benzene: 140 pm.

This is yet another fairly sound assumption and I will show you why. Eleven bonds of 140 pm will give us the following total length: $L = 11 \times 140$ pm $= 1{,}540$ pm $= 1.540 \times 10^{-9}$ m. If you check an organic chemistry textbook you will find that the lengths of double and single bonds in conjugated molecules vary between 136 and 146 pm. When we insert these numbers, for five single and six double bonds, we get the following total length: $L(2) = 5 \times 146$ pm $+ 6 \times 136$ pm $= 1{,}546$ pm $= 1.546 \times 10^{-9}$ m. The error we have made using a single bond length, $l = 140$ pm, is less than 0.4%. I think we can live with this much imprecision. So whenever you have a similar problem use the average bond length of 140 pm. Let us summarize our strategic analysis: You have a model of 12 $\pi(p)$ electrons in a potential well of width 1.54×10^{-9} m. You may now proceed in the same way we have done in the previous problem, that is, place two by two electrons on each rung and calculate their energy.

» **Solution A– Calculation**
The energy of an electron inside a square potential well is given by the expression

$$E_n = n^2 \times h^2 / 8 m_e L^2$$

Here h is the Planck's constant, m_e the mass of an electron, L the width of the potential well, and n the quantum level, that is, the number of the rung on which the electron is perched like a bird. Let us calculate the energy of the first electron; we will place it at the lowest rung, $n = 1$. A choice of lowest quantum numbers, and therefore the *lowest energy*, is a natural one; it is known as the *ground electronic state* of a molecule (or an atom, ion, radical, ..., a particle). We will label it as E_0. This is an important concept in chemistry and physics.

Make a note: The *ground electronic state*.

Now we proceed with number crunching:

$$E(1) = 1^2 \times (6.626 \times 10^{-34} \text{ [J s]})^2 / 8 \times 9.109 \times 10^{-31} \text{ [kg]} \times (1.54 \times 10^{-9} \text{ [m]})^2$$

$$E(1) = 1^2 \times 2.540 \times 10^{-20} \text{ [J]}$$

It may look a little silly to keep writing the 1^2 factor but it will help you keep track of the things. You may place the second electron on the same rung and its energy will be the same:

$$E(2) = 1^2 \times 2.540 \times 10^{-20} \text{ [J]}$$

Notice the number 2.540×10^{-20} [J]: it is repeated for each quantum level and you should keep it. What is next? You have the 3rd, 4th, ... and, finally, 12th electron and you are going to place them into the level with $n = 2$ (two electrons), $n = 3$ (two electrons), and so on, up to and including the level with $n = 6$. So the total energy of the 12 $\pi(p)$ electrons arranged on the lowest six rungs of the well will be given as

$$E(1:12) = (1^2 + 1^2 + 2^2 + 2^2 + 3^2 + 3^2 + 4^2 + 4^2 + 5^2 + 5^2 + 6^2 + 6^2) \times 2.540 \times 10^{-20} \text{ [J]}$$

Let us again remind ourselves of one thing: when all electrons in a molecule are arranged in the lowest available levels (rungs in a ladder) we say that the molecule is in the *ground electronic state*:

$$E_0(1:12) = 182 \times 2.540 \times 10^{-20} \text{ [J]} = 4.623 \times 10^{-18} \text{ J}$$

» **Solution B – Strategy and Calculation**

The next question is, What is the HOMO–LUMO energy gap and what is the wavelength of the light corresponding to this energy difference? This needs a little explanation. Let me introduce a couple of new words. HOMO means highest occupied molecular orbital. In our case this is the highest rung on which you placed two electrons, $n = 6$. The next higher rungs, $n = 7, 8, 9, \ldots$, are unoccupied. Lowest among the unoccupied levels, the lowest unoccupied molecular orbital $= $ LUMO, is the level with $n = 7$. Now imagine that a photon of light of wavelength λ and energy E falls on the retinal molecules in your eye. The photon and the $\pi(p)$ electrons in the molecule interact – they exchange energy. The photon gives away energy and one of the electrons in retinal takes in (absorbs) this light energy and jumps into the next, higher energy level. Only the electrons in HOMO have open space above them. So one of the HOMO electrons absorbs the light energy and jumps to the empty LUMO, the level with $n = 7$. This is how we describe the absorption of light by a molecule or by an atom.

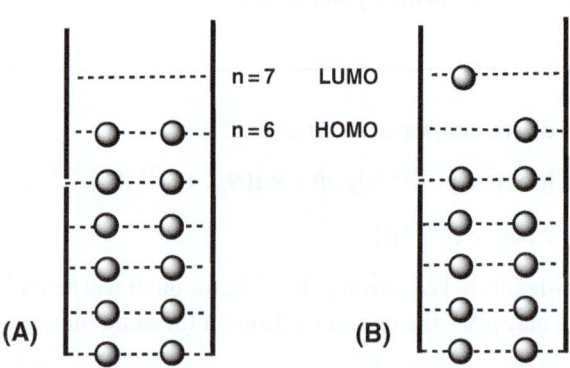

Fig. 11.4 The $\pi(p)$ electrons in the ground state retinal (A) and excited state retinal (B)

After a retinal in the ground state, (Fig. 11.4(A)), absorbs the photon energy it has more energy than it had before and we say the molecule is in *excited state*, (Fig. 11.4(B)). (There are more than one ways a molecule can get excited so this particular state should be called an "electronically excited state.") We usually put a star by a name, a formula, or another symbol we use for a particle in excited state. So for the retinal with the $\pi(p)$ electrons arranged as in (Fig. 11.4(B)) you may write (retinal)*, and it would mean a retinal molecule that just absorbed light energy and has an electron in an excited energy level, in LUMO. And what is the energy of the excited retinal, E^*? It is the same as that of the retinal in the ground state except for one electron which is now occupying the LUMO, with $n = 7$:

$$E^*(1:12) = (1^2+1^2+2^2+2^2+3^2+3^2+4^2+4^2+5^2+5^2+6^2+7^2) \times 2.540 \times 10^{-20} \text{ [J]}$$

The energy difference between E^* and E^0 is called the HOMO–LUMO gap. Like in thermodynamics, where we are considering the energy change of a chemical reaction, here we define the energy change as a difference between the excited state energy and the ground state energy, $\Delta E = E^* - E^0$. So, strictly speaking, it would be more correct to define this difference as $\Delta E = E(\text{LUMO}) - E(\text{HOMO})$, the LUMO–HOMO gap. However, someone, long time ago, named it wrongly and the name stuck. Let us see what we get for ΔE:

$$\Delta E = E^*(1:12) - E_0(1:12)$$

$$\Delta E = [(1^2+1^2+2^2+2^2+3^2+3^2+4^2+4^2+5^2+5^2+6^2+7^2) \times 2.540 \times 10^{-20}]$$
$$-[(1^2+1^2+2^2+2^2+3^2+3^2+4^2+4^2+5^2+5^2+6^2+6^2) \times 2.540 \times 10^{-20}]$$

As you can tell, many terms in the expressions for E^* and E_0 are the same and they get canceled; this is why I had asked you to keep the 1^2, 2^2, and other such terms, rather than summing them up into a single number. The remaining terms are the energy of one electron in the LUMO and the energy of one electron in the HOMO; we calculate it by subtracting just the two numbers which are different:

$$\Delta E = (7^2-6^2) \times 2.540 \times 10^{-20} \text{ [J]}$$

$$\Delta E = 3.302 \times 10^{-19} \text{ [J]}$$

This is the energy of the HOMO–LUMO gap in our model of the retinal molecule. Note that, like in thermodynamics, physical chemists often write E when in fact this is ΔE, a *difference* between two energies. And – you will calculate the HOMO–LUMO energy difference for any other molecule or atom in exactly the same way. Now what? The second part of the question (B) is, What is the wavelength of the light that corresponds in energy to the HOMO–LUMO gap in retinal? We will answer this, and all other questions of this kind, with the help of the following assumption.

» **Assumption #3**

- The energy difference between an electron in LUMO and an electron in HOMO, the HOMO–LUMO gap, equals the energy of the photon of light that caused the HOMO–LUMO electron jump.

This perhaps looks more complicated than it is. Let us just calculate the answer. First, we know the HOMO–LUMO gap, we just calculated it:

$$\Delta E(\text{LUMO}-\text{HOMO}) = 3.302 \times 10^{-19} \text{ J}$$

Next, we equate it with the energy of a photon that interacted with retinal:

$$E(\text{light}) = \Delta E(\text{LUMO}-\text{HOMO}) \qquad (11\text{-}7)$$

What we need to find out now is the wavelength of the light; we will do this by using the Planck–de Broglie's formula for energy of the photon of light:

$$E(\text{light}) = h\nu = hc/\lambda$$

$$\lambda = hc/\Delta E(\text{LUMO}-\text{HOMO}) \qquad (11\text{-}8)$$

Insert the numbers for Planck's constant, h, the speed of light, c, and the HOMO–LUMO energy difference, ΔE, and you will get

$$\lambda = 6.626 \times 10^{-34} \text{ [J s]} \times 2.998 \times 10^8 \text{ [m s}^{-1}\text{]}/3.302 \times 10^{-19} \text{ [J]}$$

$$\lambda = 601.6 \times 10^{-9} \text{m} = 602 \text{ nm}$$

This is the answer to the problem: the wavelength of the photon of light that can make a π-electron in the retinal molecule jump across the HOMO–LUMO gap is 602 nm. This is within 5% of 578 nm, one of the wavelengths to which human eye is very sensitive. Our very simple model provided us an answer within the ball-park of a correct value. Now – this was quite a problem but you have also learned few things and you will be able to tackle *any* problem of this kind; not a small achievement.

A note on HOMO and LUMO: As mentioned before the HOMO–LUMO is an impre-cise and slightly incorrect name but the concept has been so important that the name stuck and has been accepted throughout chemistry, biochemistry, physics, and materials science. It is the difference between the highest level occupied by electrons (the ground state) and the lowest unoccupied level (excited state). In molecules, as we have just shown, it (approximately) corresponds to the highest occupied molecular orbitals and lowest unoccupied molecular orbitals; in atoms it is the corresponding atomic orbitals. But the concept, as we said, extends beyond molecules, light, and orbitals. In solid state physics and materials science it is used to describe the energy needed for a bound ground state electron to "jump" to a delo-calized, unoccupied level. If the HOMO–LUMO gap is wide (high energy difference) the material has properties of an electric *insulator* (natural rubber, air). If it is narrow (low energy needed for an electron jump) the material is a *conductor* (a copper wire, aluminum foil). And yes, there are materials that come between these two classes; they are called *semiconductors* and we use them to make little switches in our com-puters, our MP3 players, and our cell phones. One may make a reasonable argument that semiconductors have been the most important material of the second half of the 20th century.

Make a note: *insulator, conductor, semiconductor.*

| Problem 11.3 | "Like the circles in my mind. . . ." |

Porphyrin, a cyclic scaffolding of carbon and nitrogen atoms, is arguably the most important biological molecule. With a centrally coordinated metal, the molecule is found in hemoglobin, as the O_2/CO_2 carrier in human blood cells, and in chloro-phyll, as the sunlight antenna in plants that collects the light energy needed for the conversion of simple inorganic molecules into nutritive carbohydrates. So the con-jugation, $\pi(p)$ electron delocalization, and the interaction with photons of light are the defining characteristics of the porphin derivatives – the porphyrins. Given that porphyrins can be simplified by a model of an 18-electron, 18-center conjugated ring (Fig. 11.5), use the model of circular potential well to calculate the HOMO–LUMO energy gap [J] and the wavelength of the exciting light [nm]. The effective radius of the ring is $r = 438$ pm.

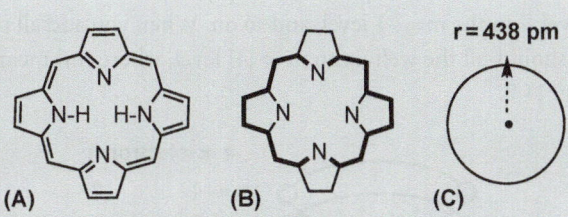

r = 438 pm

(A)   (B)   (C)

📌 **Fig. 11.5** The structure of porphyrin (A), one of its conjugated rings (B), and a circular approximation of the ring (C)

Answer: $\Delta E = 2.86 \times 10^{-19}$ J, $\lambda = 693.7$ nm

» Solution – Hints and Suggestions

We describe the energy of an electron inside a circular potential well a little differently than in the case of square potential well; it is given as

$$E = m^2 h^2 / 8 m_e \times \pi^2 r^2 \tag{11-9}$$

Here m is the quantum number (the rung), r is the radius of the circular potential well, and other symbols have their usual meaning. The quantum number m in a circular potential well deserves an additional explanation. I will show you how to fill the energy levels in a circular potential well with electrons and then you do the rest and calculate the answers to this problem.

The quantum states (rungs in a well, orbitals) in a circular system are also subject to a magnetic moment created by circling electrons, Fig. 11.6. The solutions to the equation describing this motion and the magnetic moment are whole numbers (quantum numbers) that can take positive as well as *negative* values. They start at zero which is the quantum number of the lowest level. The energies of the levels with positive and negative quantum numbers are the same. The electrons in the circular well are arranged, again, according to the two principles: (1) the minimum energy

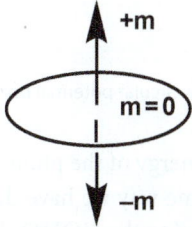

📌 **Fig. 11.6** Orientations of the magnetic moment of a rotating charged particle

and (2) the Pauli exclusion principle. Accordingly, there are 2 in the m = 0 level, 2 in the m = +1 level, 2 in the m= –1 level, and so on. When you add all porphyrin $\pi\,(p)$ electrons you should fill the well up to m = |4| level, where |m| means the absolute value of m, Fig. 11.7.

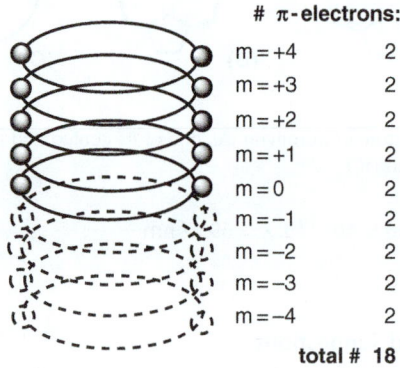

	# π-electrons:
m = +4	2
m = +3	2
m = +2	2
m = +1	2
m = 0	2
m = −1	2
m = −2	2
m = −3	2
m = −4	2
	total # 18

Fig. 11.7 Available quantum levels in a circular potential well

When porphyrin interacts with light one HOMO electron absorbs the photon energy and jumps into the LUMO level, |m| = 5; Fig. 11.8. The energy difference is, again, given as

$$\Delta E = E(\text{LUMO}) - E(\text{HOMO})$$

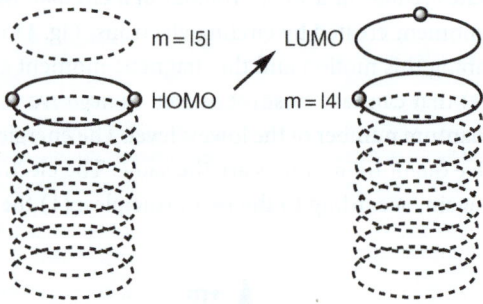

Fig. 11.8 A HOMO–LUMO jump in a circular potential box

Next, you will equate ΔE with energy of the photon of light and calculate its wavelength in nanometers, in the same way we have done in the case of retinal. When the photon energy is exactly equal to the HOMO–LUMO gap the absorption of the photon is called *resonant* transition. Like musical instruments, the electrons and photons are in resonance.

References

Units and Constants
1. Woan, G (2003) The Cambridge handbook of physics formulas. Cambridge University Press, Cambridge
2. National Institute of Standards. URL: http://physics.nist.gov/constants. Accessed July 31, 2009

Quantum Theory
3. Schrödinger E (1926) Quantisierung als Eigenwertproblem. Ann Physik, 384:361–376 and 384:489–527
4. Einstein A (1951) The advent of the quantum theory. Science, 113:82–84
5. Margenau H (1951) Conceptual foundation of the quantum theory. Science, 113:95–101
6. Sommerfeld A, Bopp, F (1951) Fifty years of quantum theory. Science, 113:85–92
7. Adler SL, Bassi A (2009) Is quantum theory exact? Science, 325:275–276

Light and Vision
8. Palczewski K, Kumasaka T, Hori, T, Behnke CA, Motoshima H, Fox BA, Le Trong I, Teller DC, Okada T, Stenkamp RE, Yamamoto M, Miyano M (2000) Crystal structure of rhodopsin: A G protein-coupled receptor. Science, 289:739–745
9. Luo D-G, Xue T, Yau K-W (2008) How vision begins: An odyssey. PNAS, 105:9855–9862
10. Goldsmith TH (1986) Interpreting trans-retinal recordings of spectral sensitivity. J Comp Physiol, 159:481–487
11. Wald G (1945) Human vision and the spectrum. Science, 101:653–658

12 Interaction of Light and Matter

12.1 UV and Visible Spectroscopy

With the last couple of problems involving HOMO–LUMO energy gaps and photon wavelengths we have laid a foundation for optical spectroscopy. Spectroscopy is a method that uses light to analyze materials. When the electromagnetic radiation is visible light we call it optical spectroscopy. Optical spectroscopy usually covers a range from 200 nm, a deep UV, to 1,000 nm, the near-infrared, NIR.

A narrower range, between about 410 nm (violet) and 720 nm (red), can be seen by the human eye and is known as *visible* light. Light, or electromagnetic radiation, extends way below the lowest visible threshold. Radiation of wavelengths of 10^{-12} m or shorter are known; the names for this type "light" are x-rays or, for even shorter wavelengths, γ-rays. This type of radiation is lethal or very harmful to living organisms. Why? Because each photon of this light can break any chemical bond in your body. On the other side, beyond the high wavelength visibility limit, 720 nm, there is infrared light; it is invisible to the eye but we feel it as heat. And yes, rattlesnakes and certain other creatures can "see" in infrared. Electromagnetic radiation extends even beyond that, all the way to the wavelength meters long – radiowaves. This, very long wavelength radiation is pervading our space all the time and is seemingly harmless (not that we can avoid it either). The whole range of light wavelengths, or more generally electromagnetic radiation, from meters to subpicometers is known as spectrum. In general, we say that all the energies – electromagnetic or other – from a certain source are a spectrum, an energy spectrum.

At the atomic and molecular level the light in this wavelength range exchanges energy with electrons. When a molecule absorbs photon energy it becomes, as we said, electronically excited. We say that light got *absorbed*. (Think of your sunglasses – they absorb light.) The time it takes for a photon of light to get absorbed is very short; likely in the attosecond (atto = 10^{-18}) to zeptosecond (zepto = 10^{-21}) range. Some physicists call it "Zeno" time, in reference to the difficult riddles associated with Zeno of Elea, an ancient Greek philosopher. Once excited, a molecule or atom relaxes fast (10^{-9} s for atoms and 10^{-12} s for molecules) releasing the absorbed energy. We say that molecule emits light or *luminesces*. There are two basic types of luminescence: *fluorescence*, which happens very fast, 10^{-10} to 10^{-14} s, after

P.-P. Ilich, *Selected Problems in Physical Chemistry*,
DOI 10.1007/978-3-642-04327-7_12, © Springer-Verlag Berlin Heidelberg 2010

the excitation and *phosphorescence* which is a light emission delayed by seconds or even hours. Both fluorescence and phosphorescence bring an electronically excited molecule back to the ground electronic state and this is why the common name for both processes is *radiative relaxation*. We do not always need light to get atoms and molecules to fluoresce: certain processing involving chemical bond breaking and making can create electronically excited molecules which then luminesce. We call this chemiluminescence or, if it occurs in living organisms (fireflies, fungi, marine planktons, and some fishes) – bioluminescence. The four basic modes of interaction of light and matter are given Fig. 12.1.

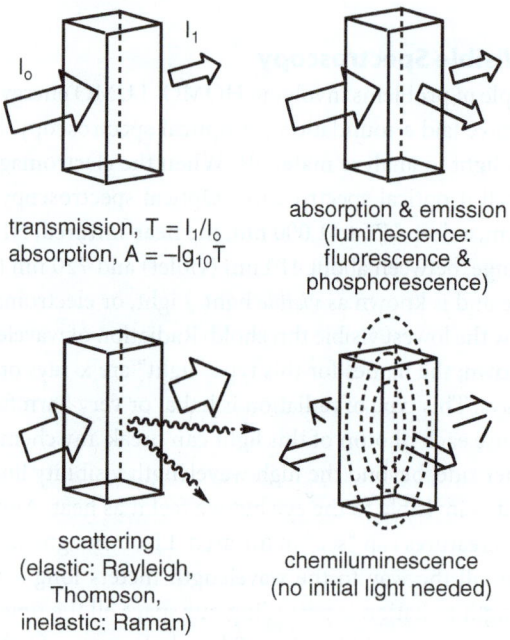

transmission, $T = I_1/I_0$
absorption, $A = -lg_{10}T$

absorption & emission
(luminescence:
fluorescence &
phosphorescence)

scattering
(elastic: Rayleigh,
Thompson,
inelastic: Raman)

chemiluminescence
(no initial light needed)

Fig. 12.1 Most common modes of interaction of light and matter

We will now look a little closer at the practical uses of *absorption* of light.

12.1.1 UV/Vis Spectrophotometry

The fact that more light is absorbed by a thicker, darker screen has been scientifically studied for almost 300 years. At first, it was noticed that the more absorbing material there is the less light will pass through (Fig. 12.2):

(1) The relation between the amount of the absorbing material and the intensity of the passed light is not linear (Fig. 12.3).

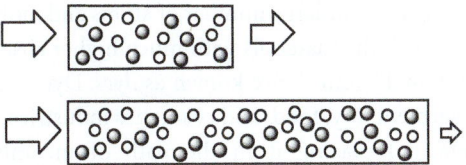

Fig. 12.2 Most of the light passes through a short cell (*top*) but most of the light gets absorbed in a long cell (*bottom*)

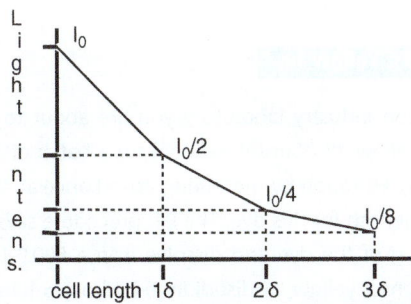

■ **Fig. 12.3** Intensity of the transmitted light falls off exponentially with the cell length

Today we use the Bouguer–Lambert–Beer relation – named in honor of the French-Swiss-German scientists, Pierre Bouguer, Johann Lambert (seemingly incorrectly), and August Beer – to describe the relation of transmitted light, absorbed light, and the amount (concentration) of the absorbing material:

$$T = I_1/I_0 \qquad (12\text{-}1)$$

$$A = \varepsilon l c \qquad (12\text{-}2)$$

$$A = -\lg(T) \qquad (12\text{-}3)$$

$$T = 10^{-A} \qquad (12\text{-}4)$$

In these formulas I_0 is the intensity of initial and I_1 of the light passed through some absorbing material ("filter"). T is transmission or *transmittance*, A is absorption or *absorbance*, ι is the thickness of the absorbing material or the lightpath, and c is the molar concentration of the light-absorbing substance. The relation between A and c is linear, which is very useful for chemical analysis, but the relation between T and c is nonlinear: the power 10 or logarithm (base 10) relation. One more thing — ε. It is the so-called *absorption coefficient* (or sometimes called extinction coefficient); ε is different for each molecule and each wavelength of light. In terms of quantum mechanics and the interaction of light with matter ε is a measure of how strongly a HOMO electron in a molecule or atom will interact with a photon of light. A and T are dimensionless, ι is given in cm, c in mol L^{-1}, and the unit for ε is L

mol^{-1} cm^{-1}. The values of ε, for large number of atoms and molecules, are between 50 and 500 L mol^{-1} cm^{-1}; the materials with much higher absorption coefficients, $\varepsilon = 50\,000$–250,000 L mol^{-1} cm^{-1}, are known as *dyes*. Dyes are used in the paint, building materials, garment, toy, food, and cosmetics industries.

Making a sample and placing it in the path of incoming light, to measure how much light is absorbed or transmitted, is known as *spectrophotometry*. Spectrophotometry is a widely used method – in teaching, industrial, medical, and research laboratories – for determining the concentration of a known substance.

Problem 12.1 | FD & C Yellow #6.

As an intern in a food industry laboratory you are about to prepare a 500.0 L solution of FD & C Yellow #6 ("Sunset Yellow") for a batch of Jell-o®. Afterward, you will check its concentration by measuring absorbance at 481 nm, the absorption maximum wavelength for this dye. The lab procedure says you should weigh and dissolve 0.6113 g of the dye and dissolve it in a 500.0 L vat of water. The molecular mass of sunset yellow, or disodium 6-hydroxy-5-[(4-sulfophenyl)azo]-2-naphthalenesulfonate, is 452.37 g mol^{-1} and its absorption coefficient at 481 nm, ε_{481}, is 5.55×10^{4} L mol^{-1} cm^{-1}. What absorbance, A_{481}, do you expect of this solution?

12

» Solution – Strategy and Calculation

A 500.00 L solution! – You better do this one right. Let us see – this is about light absorption and the method of measuring it so we should start with Beer's law:

$$A = \varepsilon l c \qquad (12\text{-}5)$$

Absorbance equals *extinction coefficient* times *pathlength* times *molar concentration*.

A good start is to make a list of things we do not know:

- Absorbance, A – we have to calculate it
- Pathlength, l
- Molar concentration, c

We are back to concentrations, it seems. For the molar concentration you need to know the number of moles and the volume. The volume is given, $V = 500.0$ L. The number of moles you will calculate by dividing the mass of the dye you weighed by its molecular mass:

$$n(\text{dye}) = \text{mass (dye)}/\text{molecular mass (dye)} = 0.6113[\text{g}]/452.37[\text{g mol}^{-1}]$$
$$= 1.351 \times 10^{-3} \text{ mol}$$

You will get the molar concentration by dividing the number of moles of solute by the volume of solvent:

$$c[\mathrm{mol\,L^{-1}}] = 1.351 \times 10^{-3}\,\mathrm{mol}/500.0\,\mathrm{L} = 2.703 \times 10^{-6}\,\mathrm{mol\,L^{-1}}$$

Now – the pathlength, the width of the optical cuvette in which you measure absorbance. In spectrophotometry l is always 1 cm, unless specified otherwise. It appears you have everything you need to calculate the absorbance, A:

$$A = (5.55 \times 10^4\,\mathrm{L\,mol^{-1}\,cm^{-1}}) \times (1\,\mathrm{cm}) \times (2.703 \times 10^{-6}\,\mathrm{mol\,L^{-1}}) = 1.499 \times 10^{-1}$$

Note: Absorbance is a unitless property. So if you have done the job right the absorption of the sample will read 0.15 when you set the spectrophotometer at a wavelength 481 nm. This example describes pretty much everything that has to be known about spectrophotometry. Always stick to Beer's law.

Problem 12.2 │ A genetically modified chicken soup.

You are working in a medical research laboratory and are using a spectrophotometer to check the quality of genetically engineered samples. In an experiment you take two milliliters of 5 mmol solution of pentapeptide, Ala-Ala-Trp-Lys-Gly, $\varepsilon_{280}(\mathrm{Trp}) = 1{,}660\,\mathrm{L\,mol^{-1}\,cm^{-1}}$, and add it to 10 mL of 2 mmol solution of its mutant analogue, Ala-Ala-Tyr-Lys-Gly, $\varepsilon_{280}(\mathrm{Tyr}) = 444\,\mathrm{L\,mol^{-1}\,cm^{-1}}$. What is going to be the transmittance (% T) of this mixture at 280 nm, if measured in a 1 mm cuvette?

Answer $T_{280} = 61.3\%$

❱ Solution – Hints and Suggestions

This is yet another problem you may encounter, for example, in a medical or biochemical lab, when using a spectrophotometer to measure the concentration of known substances. This question here is more of analytical chemistry than physical chemistry type. I would do it in the following way:

1. Notice that you are adding two solution of different concentrations, as both concentrations and volumes are known you should use them to calculate the final volume and the final concentration of both peptides, Ala-Ala-Trp-Lys-Gly and Ala-Ala-Tyr-Lys-Gly. How do you go about it?

   a. It is easy to calculate the final volume, you add the volume of the first solution, $v_1 = 2$ mL to the volume of the second solution, $v_2 = 10$ mL and get $V_{\mathrm{tot}} = 12$ mL.

b. A molar concentration is the number of moles of solute per final volume. What is the number of moles of the first peptide, Ala-Ala-Trp-Lys-Gly? You will find this from the initial concentration: $c_1(\text{init}) = 5$ mmol or 5.0×10^{-3} mol L^{-1}. Only you do not have 1 L but 2 mL only; this is $2/1,000 = 0.002$ L. To calculate the number of moles of Ala-Ala-Trp-Lys-Gly in 2 mL of the solution you will multiply its concentration, 5.0×10^{-3} mol L^{-1} by its volume 0.002 L to get $n_1 = 1.00 \times 10^{-5}$ mol. The final concentration of the first peptide, Ala-Ala-Trp-Lys-Gly, will be obtained when you add together two solutions. You will calculate this by dividing the number of moles of the first peptide, 1.00×10^{-5} mol, by the final volume, $V_{\text{tot}} = 0.012$ L. It should be $c_1(\text{fin}) = 1.00 \times 10^{-5}$ mol$/0.012$ L $= 8.33 \times 10^{-4}$ mol L^{-1}. As expected, it is smaller than the initial concentration, $c_1(\text{init}) = 5.0 \times 10^{-3}$ mol L^{-1}, because you diluted the first solution from 2 to 12 mL.

2. You will now use this concentration to calculate how much light will be absorbed by the first peptide, $A_1 = c_1 \times l \times \varepsilon_1 = 8.33 \times 10^{-4} \times 0.1 \times 1660$. Notice "0.1" in the equation instead of 1 because the pathlength in this experiment, that is the width of the cuvette you are using to do the spectrophotometric measurement, is 1 mm, not the standard 1 cm.

3. Now you will repeat the same procedure for the second solution:

a. First, you will calculate the final concentration of the second peptide, Ala-Ala-Tyr-Lys-Gly. The final volume will of course be the same as above, $V_{\text{tot}} = 12$ mL.

b. You will find the number of moles of the second peptide, Ala-Ala-Tyr-Lys-Gly, by multiplying the concentration, 2 mmol $= 2.0 \times 10^{-3}$ mol L^{-1} by the initial volume, $v_2 = 10$ mL $= 0.010$ L. Once you find the number of moles, n_2, you will divide it by the final volume, $V_{\text{tot}} = 12$ mL, to get the concentration of the second peptide in the mixture of two solutions.

4. Your will then use this second concentration, $c_2(\text{fin})$, to calculate the absorbance due to the second peptide, Ala-Ala-Tyr-Lys-Gly, using the formula, $A_2 = c_2 \times l \times \varepsilon_2 = c_2(\text{fin}) \times 0.1 \times 444$.

5. You will add the two absorbances, $A_{\text{tot}} = A_1 + A_2$, and calculate the transmittance, $T = 10^{-A_{\text{tot}}}$. Multiply this number by 100 to get % T. Done.

This is quite a problem, isn't it? But this is about as complicated as spectrophotometry and the application of the Beer's law can get. In fact, it is, again, all about *concentrations*. Once in chemistry – you just cannot escape concentrations!

12.2 Vibrational Spectroscopy

One of the major ideas in quantum mechanics is that you can know things only within certain limits. (Do I really need quantum mechanics to tell me this?)

Specifically, a statement known as the principle of uncertainty – or Heisenberg's Uncertainty Principle – tells us that once you know the interaction energy of two atoms in a molecule you cannot know their positions very accurately. This relation, named in honor of the German physicist Werner Heisenberg, can be expressed in the following way:

$$\Delta x \Delta E > h/2\pi \qquad (12\text{-}6)$$

Here Δx is the range of the atomic positions along x-axis (or any other axis we choose) and ΔE gives the limits of energy accuracy. So if I know that the energy holding together a C and an H atom in a molecule is exactly 5.585391×10^{-20} J then the best I can know the position of, for example, H atom, is no better than $\Delta x = 1.8 \times 10^{-15}$ m. Now, this may not strike you as a news one should worry about but when you put this number on the scale of chemical bond lengths, for example, $d(\text{C–H}) = 109.6 \times 10^{-12}$ m, or the diameter of a proton $d(\text{proton}) = 0.877 \times 10^{-15}$ m, then you see that no matter how hard you try you may not know the position of an atom in a molecule to better than few parts of a percent. So – an atom in a molecule is not standing still at a position fixed in space but is slightly, and constantly, oscillating. The rate of oscillation – vibration is the word preferred by physical chemists – is 10^{12} to 10^{14} s^{-1}, about trillion to hundred trillion times a second; pretty fast.

Atoms in a molecule vibrate around their equilibrium positions but not equally to the left and to the right; they are like a pendulum which is slightly pulling to one side. For each two atoms in a molecule this pendulum offset is different. Over the many years we have been studying atomic vibrations and have learned that this offset is often small and can be ignored; we assume that atoms behave almost like perfect *penduli*. This way of looking at atomic motions in a molecule is known as *harmonic vibrations* and it greatly simplifies the math we need to describe atomic vibrations (Fig. 12.4). The energy of vibration of two atoms in a molecule, for example, carbon and oxygen in carbon monoxide molecule, is given as

$$E_v = (v + 1/2)hc\sigma \qquad (12\text{-}7)$$

Here v is the vibrational quantum number, h and c are of course Planck's constant and speed of light, in this order, and σ is something new. People who measure the motions of atoms in molecules call it *wavenumber*; its unit is a reciprocal centimeter, cm^{-1}. The range of observed atomic vibrations in most molecules is typically between 250 and 3650 cm^{-1}. The vibrational quantum number v takes positive integer values: 0, 1, 2, 3,..., similar to the quantum number n in the *infinite potential well* model. Note that when $v = 0$ vibrational energy does not go to zero, $E_0 = 1/2 \times hc\sigma$; this is known as zero potential energy, ZPE.

Make a note: *ZPE – zero potential (vibrational) energy, $E_0 = 1/2 \times hc\sigma$*

Vibrational energy is related to the temperature of a molecule; the hotter the molecule the higher the vibrational quantum number v. Unlike the infinite potential well, a potential well containing the vibrational energy levels of two atoms *can* be escaped: when v gets really high so does the vibrational energy and the two vibrating atoms fall apart. A chemical bond is broken and the molecule has dissociated; we are no more concerned with atomic vibrations.

Problem 12.3 │ Stretching a rubber band.

According to harmonic approximation the ^{1}H–^{35}Cl molecule vibrates with a force constant $k = 480.6$ [N m^{-1}]. Assuming that the harmonic model holds for this molecule, and keeping in mind that the average H–Cl bond energy is 451 kJ mol^{-1}, predict the vibrational quantum level at which the H–Cl molecule will dissociate.

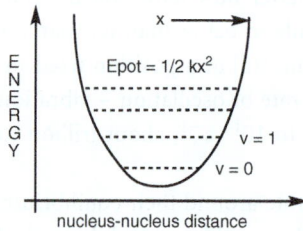

Fig. 12.4 The energy in a harmonic well depends on how far apart the atoms are

» Solution – Strategy and Calculation

The potential energy of a harmonic pendulum is given as

$$E_{pot} = 1/2 \times (k \times x^2) \qquad (12\text{-}8)$$

Here x is the distance from the center and k is the so-called force constant; it is proportional to the bond strength. You will use the force constant to calculate the vibrational frequency, v, using the following formula:

$$v = (1/2\pi) \times (k/\mu)^{1/2} \qquad (12\text{-}9)$$

In the formula above μ is the so-called reduced mass; for a diatomic molecule μ is calculated according to the formula

$$\mu = m_1 \times m_2/(m_1 + m_2) \qquad (12\text{-}10)$$

Here, m_1 and m_2 are the masses of the two atoms in the molecule; the same rule applies to 3-atomic and any-atom-number molecules, only the formulas get more complicated. For ^{1}H–^{35}Cl you will get the reduced mass using the atomic mass unit, amu, 1.6605×10^{-27} kg:

$$\mu = (1.0)(35.0)/(1.0 + 35.0)\,\text{amu} = 0.9722\,\text{amu} = 0.9722 \times 1.6605 \times 10^{-27}\,\text{kg}$$
$$= 1.614 \times 10^{-27}\,\text{kg}$$

Now you have all you need to calculate the vibrational frequency, ν:

$$\nu = (2\pi)^{-1} \times (480.6[\text{N m}^{-1}]/1.614 \times 10^{-27}[\text{kg}])^{1/2} = 0.1591 \times (2.978 \times 10^{29})^{1/2}$$
$$[\text{m kg s}^{-2}m^{-1}/\text{kg}]^{1/2} = 8.684 \times 10^{13}\,\text{s}^{-1}$$

From here you can calculate the wavenumber, σ, using the relations

$$E = h\nu = hc\sigma \qquad \text{and} \quad \sigma = \nu/c$$

When you insert the numbers for the frequency – just calculated – and the speed of light, c, you get

$$\sigma = 8.684 \times 10^{13}\,\text{s}^{-1}/2.998 \times 10^{8}[\text{m s}^{-1}] = 2.897 \times 10^{5}\,\text{m}^{-1}$$

The traditional unit for wavenumber is not m^{-1} but cm^{-1}. The conversion between reciprocal meters and reciprocal centimeters may seem confusing at first and it may help you to think this way: there are 100 cm in 1 m but when you invert both units you will have to invert their relation too, so there are 100 m^{-1} in 1 cm^{-1}. (Like other reciprocal quantities, e.g., volume^{-1}, temperature^{-1}, this is something we do not have a good intuitive feel for.) You will *divide* the wavenumber in m^{-1} by 100 to get the result in cm^{-1}

$$\sigma = 2.897 \times 10^{5}\,\text{m}^{-1}/100[cm^{-1}/m^{-1}] = 2{,}897\,\text{cm}^{-1}$$

Now that you know the vibrational wavenumber, σ, you will use it to calculate the energy of quantum mechanical oscillator, using the relation for the quantum mechanical energy of vibration:

$$E_\nu = (\nu + 1/2)hc\sigma$$

When you equate the vibrational energy with the energy of a chemical bond you will get ν, the quantum number corresponding to the vibrational state at which H and Cl fall apart. But — be careful! – you have a problem here similar to the one we encountered with UV light: you are about to compare a single apple with a freight train full of oranges. The vibrational energy given by $E_\nu = (1/2 + \nu)\,h\,c\,\sigma$ is given for *one molecule* only. The bond energy, $E_{\text{bond}}(\text{HCl})$, is given for *one mole*, that is, the Avogadro's number of HCl molecules. You will have to divide the $E_{\text{bond}}(\text{HCl})$ by the Avogadro's number before you equate it to the vibrational energy of a single molecule:

$$E_{\text{bond}}(\text{HCl})/N_A = 451 \times 10^{3}\,\text{J mol}^{-1}/6.022 \times 10^{23}\,\text{mol}^{-1} = 7.489 \times 10^{-19}\,\text{J}$$

Now set the expression for a quantum oscillator, E_ν, and solve it for ν, the vibrational quantum number:

$$(1/2 + \nu)hc\sigma = E_\nu$$
$$\nu = (E_\nu - 1/2 \times hc\sigma)/hc\sigma \qquad\qquad (12\text{-}11)$$

Do we know what E_v is? Yes — we will assume it is the same as the energy of one HCl bond, $E_{bond}(HCl)/N_A$. I suggest you make this calculation a little more transparent by first evaluating the $h\,c\,\sigma$ term which appears in both the numerator and the denominator:

$$hc\sigma = 6.626 \times 10^{-34}\,[\text{J s}] \times 2.998 \times 10^8\,[\text{m s}^{-1}] \times 2897\,[\text{cm}^{-1}] \times 100\,[\text{m}^{-1}/\text{cm}^{-1}]$$

$$hc\sigma = 5.755 \times 10^{-20}\,\text{J}$$

Note the conversion factor from cm^{-1}, the unit vibrational energies, are reported, to m^{-1}, the unit we need to use in calculations. Now it is a straightforward exercise to calculate the vibrational quantum number v:

$$v = (7.489 \times 10^{-19}\,\text{J} - 1/2 \times 5.755 \times 10^{-20}\,\text{J})/5.755 \times 10^{-20}\,\text{J}$$

$$v = 12.51$$

Quantum numbers are whole numbers, or *integers*, so we will round the result up to the nearest integer, $v = 13$.

Comment: According to the harmonic well model the $^1\text{H}^{35}$–Cl molecule, vibrating at 2897 cm^{-1}, will dissociate when it reaches vibrational quantum level, $v = 13$. Given that a harmonic well is not a good approximation for higher level vibrations – the actual potential curve becomes wider and the level spacing narrower and we call it an *anharmonic* well – it is likely that this number is higher.

12.2.1 Isotopic Effects in Molecular Vibrations

The vibrational wavenumber, $\sigma = 2897\,\text{cm}^{-1}$, calculated above is for a $^1\text{H}-^{35}\text{Cl}$ molecule. But protium, ^1H, is not the only hydrogen isotope found in nature; there is deuterium, ^2H (D), with a mass equal to two atomic units, and tritium, ^3H (T), with atomic mass of 3 amu. The hydrogen atoms found in nature are a mixture of all three isotopes and, like in any mixture, its properties – and mass – are the average value, $m(H) = 1.0078$ amu. Likewise, chlorine in nature is a mixture of about 75% of isotope ^{35}Cl and about 25% of isotope ^{37}Cl, so the average mass of Cl atoms is 35.453 amu or that of Cl_2 molecules twice this. How does this affect molecular vibrations and vibrational spectra?

If you think in terms of large "classical" bodies – and these are the only bodies that we know of and can think about – then a heavier pendulum will swing slower. The same effect is felt in the atomic world described by the quantum mechanical laws and formulas. The change of vibrational wavenumber in case of atoms of different masses can be calculated using the following equation:

$$\sigma = (1/2\pi c) \times (k/\mu)^{1/2}$$

It has been observed and calculated that the force constant (the spring resistance) does not change significantly with smaller changes in mass, the isotopic effect is felt through the change in the reduced mass, μ. So if you want to calculate the vibrational wavenumber observed in natural HCl, with masses 1.0078 for H and 35.453 for Cl you will have to re-calculate μ using the average masses:

$$\mu(\mathrm{HCl}) = 1.0078 \times 35.453/(1.0078 + 35.453)\,\mathrm{amu} = 9.7994 \times 10^{-1}\,\mathrm{amu}$$
$$= 1.6272 \times 10^{-27}\,\mathrm{kg}$$

The vibrational wavenumber will then read

$$\sigma = (1/2\pi c) \times (k/\mu)^{1/2} = (1/2 \times 3.142 \times 2.998 \times 10^{-8}[\mathrm{m\,s^{-1}}])$$
$$\times(480.6[\mathrm{N\,m^{-1}}]/1.627 \times 10^{-27}[\mathrm{kg}])^{1/2}$$
$$\sigma = (1/1.884 \times 10^{9}[\mathrm{m\,s^{-1}}]) \times (2.953 \times 10^{29}[\mathrm{kg\,m\,s^{-2}\,m^{-1}}][\mathrm{kg^{-1}}])1/2$$
$$= 2.885 \times 10^{5}[\mathrm{m^{-1}}]$$

Finally, to convert it to conventional vibrational spectroscopic units, $\mathrm{cm^{-1}}$, you divide this number by 100 $[\mathrm{cm^{-1}/m^{-1}}]$, to get

$$\sigma = 2885\,\mathrm{cm^{-1}}$$

The experimentally correct value for the vibration of a sample of natural HCl is 2885.82 $\mathrm{cm^{-1}}$ so we are 99.97% correct. The vibrational wavenumber, $\sigma = 2885\,\mathrm{cm^{-1}}$, is close to but sufficiently different from the one we obtained for the pure $^1\mathrm{H}-^{35}\mathrm{Cl}$ isotopic molecule, $\sigma = 2897\,\mathrm{cm^{-1}}$, so in vibrational spectroscopy you have to pay close attention to isotopic masses. Sometimes we use these effects to mark a particular group of atoms by causing a shift in their vibrational wavenumbers. This is done by chemically synthesizing molecules with one kind of isotopes, for example, by preparing $^2\mathrm{H}-\mathrm{Cl}$. This method has been used extensively in the vibrational spectroscopic analysis of larger biological molecules with complicated vibrational spectra. The major instruments used to measure vibrational wavenumbers and strengths are infrared spectrometers and common laboratory infrared spectrometers can detect a difference in one $\mathrm{cm^{-1}}$; better and more expensive instruments go down to 0.025 $\mathrm{cm^{-1}}$. Read the next problem.

Problem 12.4 | **Light dancing – heavy dancing.**

Isotopic substitution is an often used procedure in assigning vibrational spectroscopic features but the difference in isotopic masses of the vibrating atoms has to be sufficiently large to make the vibrational shift observable. In a double isotopic replacement experiment, a proline with a $^{13}\mathrm{C}=^{18}\mathrm{O}$ carbonyl group is inserted in a polypeptide using molecular recombinant techniques. Given that in the naturally found proline the $^{12}\mathrm{C}=^{16}\mathrm{O}$ group exhibits a narrow vibrational feature at 1,711 $\mathrm{cm^{-1}}$, calculate the shift of the carbonyl stretching vibration for the $^{13}\mathrm{C}=^{18}\mathrm{O}$ proline isotopic congener. Will you be able to observe this effect if the smallest wavenumber shift you can measure on your IR spectrometer is 2 $\mathrm{cm^{-1}}$ or larger?

» Solution – Strategy

When you cut through the technical words – double isotopic, molecular recombinant, isotopic congener – you will see that this is not so difficult a problem. It involves a polypeptide, with proline, Pro, as one of its amino acids. The vibration (stretching) of the C=O group, containing the most abundant isotopes, ^{12}C and ^{16}O, is at $\sigma(^{12}C{=}^{16}O) = 1{,}711$ cm^{-1}, Fig. 12.5. The question is what is going to be the wavenumber of the same vibration if the carbon and oxygen have been replaced by heavier isotopes, $\sigma(^{13}C{=}^{18}O)$, Fig. 12.6.

◻ **Fig. 12.5** Proline with normal isotope abundance

◻ **Fig. 12.6** Proline with ^{13}C and ^{18}O isotopes built in by synthesis

We can solve this type of problem if we make the following two assumptions.

» Assumptions #1 and #2

- The force constant, k, is about the same in ^{13}C=^{18}O as in the natural ^{12}C=^{16}O proline.
- The isotopic changes in the carbonyl group will have little effect on the vibrations of other atoms in proline.

The first assumption has been shown to hold fairly well for many different pairs of atoms. The second assumption is helping us ignore the fact that the atomic vibrations in proline are more complicated than the "bond stretching" vibrations in a single CO molecule, for example. As many experimental and theoretical studies have shown this is a fairly solid assumption for the carbonyl group. So the only variable you have to take into account are the atomic masses for C and O. You should first calculate the reduced masses of both carbonyl groups, ^{12}C=^{16}O and ^{13}C=^{18}O, and then you can use the expression that relates σ, k, and μ:

$$\sigma = (1/2\pi c) \times (k/\mu)^{1/2}$$

I suggest that, before you start punching the keys on your calculator you first write down one such expression for $^{12}C{=}^{16}O$ and then another for $^{13}C{=}^{18}O$ and compare them. I think you will be able to cancel several terms and end up with a really simple calculation.

» Solution – Calculation

First, let us determine the reduced masses for the $^{12}C{=}^{16}O$ and $^{13}C{=}^{18}O$ groups

$$\mu(12,16) = 12 \times 16/(12 + 16) = 6.857\,\text{amu} = 6.857 \times 1.6605 \times 10^{-27}\,\text{kg}$$
$$= 1.139 \times 10^{-26}\,\text{kg}$$

$$\mu(13,18) = 13 \times 18/(13 + 18) = 7.548\,\text{amu} = 7.548 \times 1.6605 \times 10^{-27}\,\text{kg}$$
$$= 1.253 \times 10^{-26}\,\text{kg}$$

Now, let us write the expressions for wavenumbers, for $\sigma(12,16)$ and $\sigma(13,18)$

$$\sigma(12,16) = (1/2\pi c) \times (k/\mu(12,16))^{1/2}$$

$$\sigma(13,18) = (1/2\pi c) \times (k/\mu(13,18))^{1/2}$$

This is a perfect case to apply *little big trick #1*: divide the second equation by the first and cancel the same terms:

$$\sigma(13,18)/\sigma(12,16) = (1/2\pi c) \times (k/\mu(13,18))^{1/2}/(1/2\pi c) \times (k/\mu(12,16))^{1/2}$$

You may cancel everything on the right side except $\mu(13,18)^{1/2}$ and $\mu(12,16)^{1/2}$; this will give you

$$\sigma(13,18)/\sigma(12,16) = \mu(12,16)^{1/2}/\mu(13,18)^{1/2}$$

You know everything but $\sigma(13,18)$, which is what we are calculating. Insert the numbers for $\sigma(12,16)$, $\mu(12,16)$, and $\mu(13,18)$, punch a few keys on your calculator, and you will get

$$\sigma(13,18) = 1{,}711[\text{cm}^{-1}] \times (1.139 \times 10^{-26}\,\text{kg}/1.253 \times 10^{-26}\,\text{kg})^{1/2}$$

$$\sigma(13,18) = 1{,}711\,\text{cm}^{-1} \times 0.9531 = 1{,}631\,\text{cm}^{-1}$$

So when the $^{12}C{=}^{16}O$ isotopes (most abundant) in a proline molecule in a polypeptide sample are replaced by their isotopes, $^{13}C{=}^{18}O$, the carbon–oxygen vibration will shift from 1,711 and 1,631 cm^{-1}. This amounts to the following difference: $\Delta\sigma = 1{,}711 - 1{,}631\,\text{cm}^{-1} = 80\,\text{cm}^{-1}$. This is a large shift in vibrational spectra, much larger than the 2 cm^{-1} resolution of your spectrometer and you will be able to easily observe it in an otherwise rather complicated infrared spectrum of a polypeptide.

12.3 Nuclear Magnetism and NMR Spectroscopy

When you say magnet you probably think of a solid metallic body that sticks to a steel object, a refrigerator door for example. All magnets we know have two poles, *north* and *south*. Physicists call such objects magnetic dipoles. When you have one magnet it usually does not matter where its north pole or south pole is pointing.

It is quite different when you have two magnets: they can take two types of positions in space, four positions in total. They *attract* each other when south is oriented toward north, Fig. 12.7, or they *repel* or repulse each other when the same poles, for example, north and north, are oriented toward each other, Fig. 12.8.

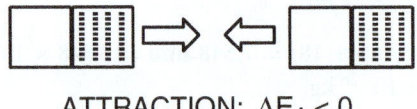

ATTRACTION: $\Delta E_A < 0$

Fig 12.7 Two magnets in attractive position; a lower energy state

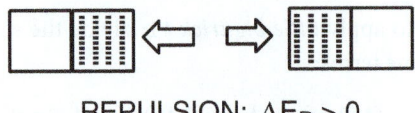

REPULSION: $\Delta E_R > 0$

Fig 12.8 Two magnets in repulsive position; a higher energy state

The total energy, or Gibbs free energy, of the two magnets after the interaction is lower when they are positioned to attract each other; we say this is a stable or the ground state. (Like the state for electrons at the bottom rungs of a potential well, or the state of a molecule vibrating at the lowest level, $v = 0$.) Two magnets, if left undisturbed, will take and retain the attraction position.

Now you may pry the two magnets apart and position one of them so that its north pole is facing the north pole of another magnet. You use the mechanical strength of your fingers to do this: you impart energy to both magnets. The two magnets are now in the repelling position and you can feel this resistance if you are holding them. The magnets you just flipped around using the strength of your fingers are in a higher energy state or in an excited state. Let us use symbols and formulas to describe what you have done:

$$\Delta E_A + \Delta E(\text{finger work}) = \Delta E_R, \quad \Delta E_R > \Delta E_A$$

The repelling state is not a stable state; it tends to give the extra energy away and return to the ground state. When this happens we say that a magnet in excited state has *relaxed*. So we can write a string of words describing these changes:

12

Ground state $+ \Delta E \rightarrow$ *Excited* state $- \Delta E \rightarrow$ *Ground* state

Now you build a large, strong magnet, set it in a laboratory, and place in its field a very small second magnet, an atom. The nucleus in the atom, say a hydrogen atom, is a charged particle seemingly fast spinning and this makes it magnetic. So here you have the same situation: two magnets interacting. At first, they will assume the lower energy position, the ground state, Fig. 12.9. You record this position. Then you impart energy to the little atomic magnet and flip it so that its north pole is facing the north pole of the laboratory magnet, Fig. 12.10. The atomic magnet is now in an excited state. How do you flip an atomic magnet into a state of higher energy? You shine a light on it, a radiation (for it is not visible light) of certain frequency and energy.

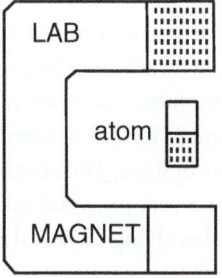

🔲 **Fig. 12.9** The laboratory and nuclear magnets in attractive position

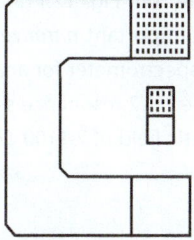

🔲 **Fig 12.10** The laboratory and nuclear magnets in repulsive position

The energy of this radiation equals the energy difference between excited state and the relaxed, ground state, of the nuclear magnet. How do we call this light energy? We call it *resonant* energy as the energy of the radiation and the energy needed to flip the nuclear magnet between the two states, the ground and excited state, are the same. They are like musical instruments in resonance, in perfect tune. So you record when the nuclear magnet absorbs the radiation and jumps into higher energy state. Now you switch this radiation off and let the excited nuclear magnet relax into the ground state; you record this relaxation. We call this record an NMR line, or an NMR signal. This whole experiment is therefore named nuclear magnetic resonance, NMR, and it can be used as a very versatile method to probe and analyze atomic nuclei in a molecule – the NMR spectroscopy. Since the higher magnetic state of the atomic nucleus is *induced* by an external action the relaxation back to the ground state is called free induction decay, FID.

Electrons, like nuclei, have magnetic moments too, and they affect and slightly change the atomic nuclear magnet. Depending on the electrons surrounding the nucleus – the number of electrons and the type of orbitals – the resonant energy of the nuclear magnet will slightly shift to higher or lower energy. For example, the resonance energy needed to excite the H nucleus in an H_2 molecule is different from the resonant energy needed to excite an H nucleus in methane, CH_4. This is what we call *chemical shift*. Knowledge of chemical shifts helps us identify different atoms and asses their bonding and non-bonding interactions in a molecule.

Also, since there are several atoms in a molecule – or several million atoms in a large biological molecule – the nearby nuclei interact with each other through their magnetic moments. This interaction, known as nuclear–nuclear spin coupling, causes splitting of the NMR resonant lines in a proton magnetic resonance spectrum into two, three, or larger number of lines. The observed chemical shifts and nuclear spin coupling provide a basis for a very powerful method for studying the nature of molecules. Here is a little riddle involving chemical shifts and magnetic fields.

12

| Problem 12.5 | Watch the (traffic) signals. |

The presence of peroxynitrite in mitochondria leads to protein nitration at the benzene ring of aminoacid tyrosine, Tyr, Fig. 12.11. Little understood, this reaction is considered to be part of the important *nitrative signaling* in cells. In a biological experiment, using an NMR spectrometer for analysis, Tyr-C2 nitration causes a 2,131.5 Hz frequency shift in the ^{13}C2 resonance signal. How large chemical shift, in ppm, is expected in a magnetic field of 7.1100 T? The ^{13}C gyromagnetic ratio is 1.0707 [2π MHz T^{-1}].

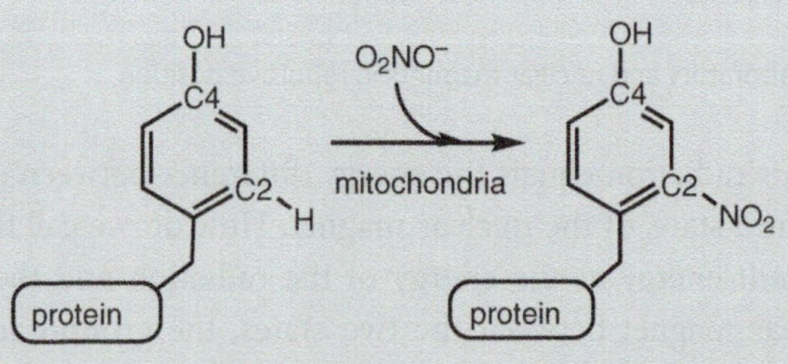

■ **Fig. 12.11** Nitration of Tyr residue in certain tissues

» Solution – Strategy

The riddle involves two relations important for understanding basic NMR spectroscopy: the strength of the magnet and the frequency of the radiation used to promote nuclear magnets to excited states. The strength of the magnet is labeled as B, $B = 7.110$ T (Tesla), and the frequency ν, used to excite a ^{13}C nucleus, is given by the following relation:

$$\nu = \gamma(C) \times B \tag{12-12}$$

Here γ or, $\gamma(C)$ for ^{13}C nucleus, is a parameter which is a property of spinning bodies with magnetic moment. It is called gyromagnetic (or, sometimes, magnetogyric) ratio. The experimentally determined value for $\gamma(^{13}C)$ is 1.0707 [2π MHz T^{-1}]. The 2π in the units comes from the relation of linear (or longitudinal) frequency, ν, and the angular frequency, $\omega = 2\pi\,\nu$, used to describe rotational motion. (Think of this: the frequency of Earth's trajectory around the Sun is one revolution per year.) The frequency obtained from the equation given above is also known as the basic frequency of the NMR spectrometer, ν_0. Once you calculate ν_0, in Hz, or s^{-1}, you will use it to calculate the chemical shift corresponding to the frequency shift caused by the chemical change on C2 in the tyrosine ring. The expression relating the measured frequency, ν, the basic frequency, ν_0, and chemical shift, δ, is given as

$$\delta[\text{ppm}] = \{(\nu - \nu_0)/\nu_0\} \times 10^6 \tag{12-13}$$

A short note on biological Tyr nitration: First, notice the labeling of tyrosine atoms: it is opposite to the labeling prescribed by organic chemistry nomenclature rules. The atom C1 is not the one with the –OH functional group, as in phenol, but the atom connected to the rest of the protein. Second, notice the position of the nitro group in nitrated tyrosine: it is *meta-* to the strongly *ortho-/para*-directing OH group. This is no mistake: nitration of tyrosine residue in certain proteins proceeds through a *radical mechanism*, following rules different than those operative in electrophilic aromatic substitution chemistry.

» Solution – Calculation

You will calculate the basic frequency by inserting the values for γ and B, as follows:

$$\nu_0 = \gamma(C) \times B$$

$$\nu_0 = 1.0707[2\pi\,\text{MHz}\,T^{-1}] \times 7.1100[T] = 76.127\,\text{MHz} = 7.6127 \times 10^7\,\text{Hz}$$

Note that if the unit for γ was given in frequency units per magnetic field units, for example, as MHz T^{-1}, instead of the 2π MHz T^{-1} as we have had, then you would need to divide the $\gamma \times B$ product by 2π, that is, $\nu_0 = \gamma(C) \times B/2\pi$. So – like in all other areas of physical chemistry – watch for units.

The difference between the measured and the basic frequency, $(\nu - \nu_0)$, is already given so you will just insert this number and calculate the chemical shift, δ:

$$\delta[\text{ppm}] = \{2131.5[\text{Hz}]/7.6127 \times 10^7 [\text{Hz}]\} \times 10^6$$

$$\delta = 28.0[\text{ppm}]$$

A comment on nuclear magnetic energies: Note the units for chemical shift: ppm – parts per million. The ratio of frequency shifts, usually given in tens, hundreds, or at most thousands of $[s^{-1}]$, and the basic frequency, usually ranging from 300 to 600 MHz and higher, would be a very small number. For this reason the factor 10^6 has been added to express chemical shifts in numbers that are easy to remember: 1.2 or 28.0. Therefore the name for it: parts per million. So the C2 in tyrosine, after nitration, will shift downfield by 28 ppm. Note that a shift toward *larger ppm* values is considered a *downfield shift* in NMR spectroscopy. Also note another important thing: The frequency of the exciting radiation ("light") is measured in megahertz, MHz, typically in the 250– 900 MHz range. This may look like a large number but, as a the next problem illustrates, it is actually millions of times smaller than the frequency of UV and visible light. The frequency of the radiation used to magnetically excite atomic nuclei is in the range of radiowaves. In fact, older type NMR spectrometers used frequencies in the 60-100 MHz range, just like your local "FM 100" radio station. This is why the excitation radiation used in NMR spectroscopy is also referred to as *radiofrequency*.

Problem 12.6 | A slogger, a smack, and a puff.

A sample of benzaldehyde isolated from *almond oil* is analyzed by three kinds of molecular spectroscopy: UV/Vis, vibrational, and NMR. In the UV/Vis spectrum a weak absorption was observed at 310 nm, due to the so-called $\pi^* \leftarrow n$ electronic transition (an electron from a non-bonding molecular orbital, n, jumps into an empty anti-bonding π^* orbital). In the vibrational spectrum, there is a strong absorption at 1,686 cm^{-1} likely due to the C=O stretching vibration. In the ^1H NMR spectrum of benzaldehyde, run at 300.1318534 MHz base frequency, a characteristic line is observed at 10.6 ppm, likely indicating the presence of the aldehyde proton. Calculate and compare these three energies by setting the UV photon energy to one.

» Solution – Calculation

(A) UV energy:

$$E_{310} = hc/\lambda$$

$$E_{310} = 6.626 \times 10^{-34}\,J\,s \times 2.998 \times 10^8\,m\,s^{-1}/310 \times 10^{-9}[m] = 6.408 \times 10^{-19}\,J$$

(B) C=O vibrational energy:

$$E_{1686} = 1/2 \times hc\sigma$$

$$E_{1686} = 1/2 \times 6.626 \times 10^{-34}\,\text{J s} \times 2.998 \times 10^8\,\text{m s}^{-1} \times 1686 \times 100[\text{m}^{-1}]$$

$$= 1.675 \times 10^{-20}\,\text{J}$$

(C) Radiofrequency energy:

$$\delta = (\nu - 300{,}131{,}853.4/300{,}131{,}853.4) \times 10^6$$

$$\nu = (10.6 \times 300{,}131{,}853.4 \times 10^{-6} + 300{,}131{,}853.4[\text{s}^{-1}]) = 300{,}135{,}034.8[\text{s}^{-1}]$$

$$E = h\nu = 6.62606896 \times 10^{-34}\,\text{J s} \times 300{,}135{,}035[\text{s}^{-1}] = 1.989 \times 10^{-25}\,\text{J}$$

Comparison of the three energies:

$$E(\text{UV}){:}E(\text{vib}){:}E(\text{NMR}) = 6.408 \times 10^{-19}\,\text{J}{:}1.675 \times 10^{-20}\,\text{J}{:}1.989 \times 10^{-25}\,\text{J}$$

$$E(\text{UV}){:}E(\text{vib}){:}E(\text{NMR})/6.408 \times 10^{-19}\,\text{J} = 1 : 0.026 : 0.00000031$$

As you can tell the light energy that can excite an "average" molecule is tens to hundreds of times higher than the lowest vibration energy; it is also several million times higher than the radiofrequency used to excite a nuclear magnet.

12.4 Level Population

The basic scenario for an interaction of light and matter is like this: a photon of light (UV, visible, infrared, radiofrequency) falls onto material and causes the orbital-bound electrons, or the diatomic oscillators, or the nuclear magnets to be excited. The energy of light, expressed as wavelength, wavenumber, or frequency, is important in this interaction. It has to be of the right magnitude in order to cause a transition, a jump, of an electron or an atomic magnet from the ground to an excited state, to cause the so-called resonant excitation. But so is the very *number of these transitions* important. Assume that you have one mole of a dye and you shine resonant light on it. In this process one photon gets absorbed by an electron in a single molecule and then it falls back to the ground state. You would most likely not notice that anything has happened. Though the methods of detection of such processes are becoming increasingly better you need a much higher number of molecules to get excited from the ground to excited state.

But the problem may not be so much with the *number of molecules* that get excited and are then allowed to relax – the detection limits will be lower and lower in the future – as the very *nature of the ground and excited states* in the material we are analyzing. In case of UV and visible light a transition from the ground to an excited electronic state is easy to accomplish: all excited levels are empty and there is a lot room for "excitation." Similar to it is the situation with vibrational levels in a molecule: most of higher vibrational states are empty. Only at a very high temperature do higher vibrational levels start getting populated by vibrationally excited molecules.

In the case of nuclear magnets the situation is very different. The excited nuclear magnetic levels have only a slightly higher energy than the ground state. As a consequence, many of the excited-state nuclear magnetic levels are already populated by nuclear magnets even at room temperature. Since there are relatively few empty excited-state levels the very interaction of the matter with radiofrequency is weak – as very few, if any, magnets get flipped up through this interaction. (It is a little bit like this: you need to take just one more course in order to graduate but the enrollment at your college has stopped; the course is full.) As a consequence, NMR spectroscopy is a method of low sensitivity. You need a lot of sample and the transition signal you are able to observe is very weak and has to be processed by sophisticated mathematical–numerical procedures to be useful for chemical analysis. This – the amount or the concentration of the sample – can become an issue in case of biological materials which are often available in small quantities and at a high cost only; think of a genetically engineered experimental drug. The following riddle should help you understand and appreciate this problem.

12

| Problem 12.7 | The upper and the lower house. |

A liquid NMR sample is placed in the magnet and excited with a frequency of 300,131,853.4 Hz. What would be the ratio, in part per million [ppm] between the excited, N_β, and relaxed, N_α, number of the proton nuclear spins at 305.00 K? How much, in ppm, and in which direction would this ratio change when you lower the sample temperature to 275.00 K?

» Solution – Strategy

A relation that expresses a ratio between two levels separated by energy E is given by the so-called Boltzmann distribution – the name we already encountered in the chapter on entropy:

$$N_\alpha/N_\beta = \exp[-E/k_\mathrm{B}T] \tag{12-14}$$

Here N_β is the number of particles in higher energy level and N_α the number of particles in the lower energy level. The constant k_B is Boltzmann constant, and T is absolute temperature. $k_\mathrm{B} = 1.3806504 \times 10^{-23}$ J K^{-1} [1]; you may think of it as the gas constant per one particle: $k_\mathrm{B} = R/N_\mathrm{A}$.

» Solution – Calculation

You will solve this problem by first calculating the energy needed for the magnetic excitation, E, by using the Planck–de Broglie's formula for the energy of electromagnetic radiation:

$$E = h\nu$$

$$E = 6.62606896 \times 10^{-34}[\text{J s}] \times 3.001318534 \times 10^{+8}[\text{s}^{-1}] = 1.98869436 \times 10^{-25}\,\text{J}$$

Insert all the values you know in the Boltzmann's distribution formula and you will get

$$N_\alpha/N_\beta = \exp\{-1.98869436 \times 10^{-25}\,\text{J}/1.3806504 \times 10^{-23}[\text{J K}^{-1}] \times 305.00[\text{K}]\}$$

For $T = 305$ [K] we have

$$N_\alpha/N_\beta = 9.9995(28) \times 10^{-1} (\text{note that the last two digits are uncertain})$$

Keeping in mind that $N_\alpha = 1 - N_\beta$, we get for N_β

$$N_\beta/(1 - N_\beta) = 0.99995(28) \text{ and} N_{\beta(305\,\text{K})} = 0.99995(28)/(1 + 0.99995(28))$$
$$= 0.49998(82)$$

Now, keep in mind that the total number of proton magnets in a molecule is 100%. If they are equally distributed between the ground and the excited state, the population of either state, N_α or N_β, will be exactly 50% or 0.5000000. The number you got for N_β is a little smaller than 1/2. Let us see how much smaller: subtract it from 1/2.

$$0.5000000 - 0.49998(82) = 0.00001(18)$$

Multiply the result by 100 and you get 0.0012% – the total availability of excited proton magnetic states at 305 K. Note that if both N_α and N_β were exactly 0.5 no transitions between the ground and excited state would be possible and the NMR spectroscopy as a method would not exist.
For $T = 275$ K you will get

$$N_\alpha/N_\beta = \exp\{-1.98869436 \times 10^{-25}\,\text{J}/1.3806504 \times 10^{-23}[\text{J K}^{-1}] \times 275.00[\text{K}]\}$$
$$= 9.9994(76)$$

And the population fraction of higher energy levels at $T = 275$ K is obtained from the following expression:

$$N_{\beta(275\,\text{K})} = 0.999994(76)/(1 + 0.99994(76)) = 0.49986(91)$$

This number is still very close to 1/2 but a little less so, that is, 0.50000 – 0.49986 (91) = 0.00001309 = 0.001309%. The difference between the two populations, at 305 and 275 K, multiplied by a million and expressed in parts per million, ppm, is given as

$$N_{\beta(305\,\text{K})} - N_{\beta(275\,\text{K})} = |(0.4999882 - 0.4998691) \times 10^6| = 119.1[\text{ppm}]$$
$$\approx 120\,\text{ppm}$$

Given the uncertainty in our calculations the result is only very approximate; still it points clearly in one direction. By lowering the temperature of the sample by

thirty degrees Kelvin you will increase the availability of the excited proton magnetic states by over one hundred parts per million. A very small change but still a
definite improvement of the experimental conditions. I hope you understand now
a little better the problem with the low sensitivity of the NMR spectroscopy. And –
keep in mind that the method of Boltzmann distribution is not meant for the atomic
magnetic levels only: it works equally well on populations of college students, liberal
voters, or spawning salmon.

12.5 Down-Conversion of Photon Energy

In spectroscopy we go to great lengths to engineer instruments which shine light of
certain wavelength (or energy) only: 200–1000 nm for the UV/Vis spectroscopy,
500–4000 cm^{-1} for vibrational spectroscopy, 100–1,000 MHz for NMR spectroscopy, or some other energy range for another type of spectroscopy. Outside
the lab, we and everything around us is subjected to all kinds of electromagnetic
radiations. (Not to mention the Solar System and Universe which are pervaded
with all kinds of electromagnetic radiation all the time.) Some of this radiation
is so hard – meaning, high energy – that it will not only excite one or two electrons but will break chemical bonds. When a chemical reaction is caused by light we
talk about *photochemistry*. This is the kind of electromagnetic radiation we humans
should avoid by all means. Softer radiation will, instead of exciting one electron
from HOMO to LUMO, knock it out of the atomic or molecular orbital and will
turn the molecule (or atom) into an ion; we call this process *photoionization*. Even
a relatively low energy visible light can cause photoionization. For the most part we
humans are used to visible light and this type of radiation will only get reflected,
scattered, or absorbed. A photon of light, no matter how hard, will also cause atoms
in a molecule to start vibrating faster. Like the way bigger (body trunk) and smaller
(legs and arms) motions of our bodies are connected, electronic and vibrational
excitations in a molecule are connected too; in physics and chemistry we say –
"coupled." This happens all the time and we say that a photon of light dissipates –
trickles down – into smaller energy packets. We will do one problem of each of these
three phenomena: photochemistry, photoionization, and electronic–vibrational
coupling.

Problem 12.8 | Avoid too much tanning.

The mid-range wavelength of the lamps used in sun-tanning beds is 360 nm. (A)
How much energy in Joules per photon does this light correspond to? (B) How
much is this in electronvolts, eV? (C) What will be the energy, in J/photon of light
half this wavelength? (D) How does that energy compare to 146 kcal mol^{-1}, the
average C=C bond enthalpy?

» Solution – Strategy and Calculation

(A) Calculating the energy of photons of light is quite simple; you use the Planck–de Broglie's formula:

$$E(\text{photon of light}) = hf \quad \text{or} \quad E = h\nu$$

Here f or ν is frequency and h is, of course, Planck's constant, $h = 6.626\,068\,96 \times 10^{-34}$ J s [2]. When wavelength, rather than frequency, is known you use the relation between the light frequency and wavelength $c = \lambda \times \nu$, where c is the speed of light, 2.99792458×10^8 m s^{-1} in vacuum [2]. So you write for the photon energy expressed by wavelength

$$E = hc/\lambda$$

You will use this formula to find the energy of light with wavelength of 360 nm. The n in nm is of course nano $= 10^{-9}$ part. So you get

$$E_{360} = 6.626 \times 10^{-34}\,\text{J s} \times 2.998 \times 10^8\,\text{m s}^{-1}/360 \times 10^{-9}\,\text{m} = 5.518 \times 10^{-19}\,\text{J}$$

Note again the order of magnitude, 10^{-19} J, so you are likely on the right track.

(B) Electronvolt, eV, is an old unit physicists have been using; it gives simple, small, easy to handle numbers. To convert an energy given in joules to electronvolts you should always think of this relation: 1 volt (electric potential) × 1 coulomb (electric charge) = 1 joule:

$$1\,\text{V} \times 1\,\text{C} = 1\,\text{J}$$

One electronvolt is the energy in joules that corresponds to an elementary charge in a potential field of one volt. How much is an elementary charge? It equals the charge of one single electron, $q(e) = 1.602176487 \times 10^{-19}$ C [2]. So – divide the energy in joules by elementary charge in coulombs and you will get the energy in electronvolts:

$$E_{360} = 5.518 \times 10^{-19}\,\text{J}/1.602 \times 10^{-19}\,\text{C} = 3.444\,\text{eV}$$

You see? The number 3.444 is much easier to remember and handle than 5.518×10^{-19}. This is why electronvolt has been used for so long time (and still is, among high-energy physicists).

(C) Half of the wavelength $\lambda = 360$ nm is 180 nm. Insert this number in the formula for energy and you will get

$$E_{180} = 6.626 \times 10^{-34}\,[\text{J s}] \times 2.998 \times 10^8\,[\text{m s}^{-1}]/180 \times 10^{-9}\,[\text{m}] = 1.104 \times 10^{-18}\,\text{J}$$

Ten to minus eighteen is ten times bigger than ten to minus nineteen so you see the relation between energy and wavelength:

The shorter the wavelength the higher the energy of the photon.

(D) The question here is, How does the energy of the light at $\lambda = 180$ nm, that is, $E_{180} = 1.104 \times 10^{-18}$ J, compare with 146 kcal mol^{-1}, the energy (enthalpy) of C=C bonds. It is important to realize the following: we are asked here to compare what seem to be apples and oranges. Actually, it is more like an apple and a freight train of oranges.

Notice the following: $E_{180} = 1.104 \times 10^{-18}$ J is the energy of *one photon* of light. $E(C=C) = 146$ kcal mol^{-1} is the energy of *one mole* of C=C bonds. So on one side you have 1 particle, 1 photon, and on the other side you have N_A, the Avogadro's number of C=C bonds. In order to equate these two values you have to bring them to the same scale, that is, (1) either you divide the $E(C=C)$ by the Avogadro's number or (2) you multiply E_{180} by the Avogadro's number. Let us multiply the photon energy by N_A:

$$E_{180}(\text{mol}) = 1.104 \times 10^{-18} \text{ J} \times 6.022 \times 10^{23} \text{ mol}^{-1} = 6.648 \times 10^5 \text{ J} = 665 \text{ kJmol}^{-1}$$

Now you will convert the C=C bond enthalpy, given in kcal mol^{-1}, to kJ mol^{-1}, by multiplying it with by 4.187 (the old – about 60 or so years – conversion factor, still found in much literature, is 4.184):

$$E(\text{mole C} = \text{C}) = 146 \text{ kcal mol}^{-1} \times 4.187 \text{ kJ kcal}^{-1} = 611 \text{ kJ mol}^{-1}$$

Now you can make a sensible comparison: the bond energy is 611 kJ mol^{-1} and the light energy is almost 665 kJ mol^{-1}. Divide these two numbers and you will get the answer:

$$E(\text{mole photons})/E(\text{mole C} = \text{C}) = 665 \text{ kJ mol}^{-1}/611 \text{ kJ mol}^{-1} = 1.088 \approx 109\%$$

If asked whether light of 180 nm can break a C=C double bond you should answer positively: the light energy is almost 9% higher than the bond energy, quite enough to start a photochemical reaction. This type of question – a comparison of light (or energy of an electromagnetic radiation) and chemical bond or some other property on macroscale – is encountered often, even in everyday life, and you should be able to answer it easily and with a good number.

| Problem 12.9 | Photo-electrochemical-kinetic smorgasbord: where O_2 is made. |

The 21% of dioxygen we are enjoying in the Earth's atmosphere is of plant (and also blue algae and cyanobacteria) origin. There, the many – meticulously chosen and carefully put together [3–6] – molecules of the photosynthetic system 1 and photosynthetic system 2, PSI and PSII, respectively, use sunlight and water to make dioxygen and glucose – so we can breathe and feed on. We still do not quite understand how this is done but we do know that oxidation of water is the driving reaction. The molecular engine in PSII, known as the oxygen evolving complex, EOC, takes four strokes of light ($\lambda_{max} = 680$ nm) and two molecules of water to make one dioxygen, O_2. (Along with this two 1,4-cyclohexa-2,5-diene-dione-type molecules, plastoquinone, PQ, are reduced and four protons are moved from one to another region of the photosynthetic cell.) The engine starts at the so-called stage S_0 and after absorbing the first photon – by the so-called light-harvesting complex (LHC1 and LHC2) in a photosynthetic organism – it needs 30 µs to shift to the next stage, S_1; after absorbing the second photon it is 70 µs of photoelectrochemistry to reach the stage S_2, then third photon and 190 µs brings it to the stage S_3, and fourth photon and 200 µs to reach the last stage, S_4. This is followed by a 1.1 ms long dark stage which completes the full cycle. Then the reaction starts anew. The photoelectrochemical reaction is given by the following scheme [6]:

$$P_{680} + 4h\nu(\lambda = 680\,\text{nm}) + 2\,H_2O + 2PQ + 4H^+(\text{stroma}) \rightarrow O_2 + 2\,PQH_2 \\ + 4\,H^+(\text{lumen}) \qquad (12\text{-}15)$$

Use these data, (1) the energy of a photon, (2) the number of photons per reaction cycle, (3) the standard reduction potential of dioxygen, $E^{\ominus} = 1.229$ V (pH 0), and (4) the total time needed for a reaction cycle to calculate the following: (A) the *apparent efficiency* of the "PSII engine" at optimal conditions (pH 5) and (B) the reaction rate constant [s^{-1}].

A comment: "Stroma" and "lumen" are different regions in a so-called thylakoid cell, the one involved in photosynthesis; the previous equation implies that a *transport* of four protons occurs at each reaction cycle.

» Solution – Strategy and Calculation

I hope you agree with me that this is a good problem to end (well, almost) this book as it makes use of several different topics we have treated separately in the past: photons, electrons, thermodynamics, kinetics. If you have followed me through most of the problems so far you are well prepared to tackle this one. As we carry on you will also realize this is by no means a difficult problem – just diverse. It is a problem from real life (slightly simplified) and like in real life it gives you several pieces of information and it asks more than one question. Let us restate the questions, one by one: (A) Calculate the apparent efficiency of the "PSII engine" at optimal conditions (pH 5). By efficiency, in this and pretty much every other case, is meant a *ratio* of the energy supplied and the energy outputted.

What is the supplied energy? The energy of light, that is,

- The energy of four photons received in one reaction cycle, $E = 4 \times hc/\lambda$

What is the outputted energy?

- The energy, that is, ΔG(reaction) of the oxidation of $2H_2O$ into O_2

Now the efficiency will be given as

Efficiency $= \Delta G(\text{reaction})/E(\text{light, mol})$

So let us translate this into formulas and numbers. First – the energy of one photon of light at 680 nm:

$$E(680\,\text{nm}) = hc/\lambda = 6.626 \times 10^{-34}[\text{J s}] \times 2.998 \times 10^{8}[\text{m s}^{-1}]/680 \times 10^{-9}[\text{m}]$$

$$E(680\,\text{nm}) = 2.921 \times 10^{-19}\text{J photon}^{-1}$$

You will need to convert the light energy from particles (photons) to moles, in order to be able to compare it to the electrochemical potentials and ΔG of the reaction. The energy of four photons, converted to moles — that is, the energy of four *moles of photons* — will be given by

$$E(\text{light, mole}) = E(680\,\text{nm}) \times N_A = 4 \times 2.921 \times 10^{-19}[\text{J}] \times 6.022 \times 10^{23}[\text{mol}^{-1}]$$

$$E(\text{light, mole}) = 7.037 \times 10^{5}\,\text{J mol}^{-1} = 703.7\,\text{kJ mol}^{-1}$$

Now – the energy used by the PSII system: it is the energy needed to create one molecule of O_2 through oxidation of two molecules of H_2O. This is an electrochemical–thermodynamic process so we will carry out all calculations in molar quantities: one mole of O_2 is formed from two moles of H_2O. Previously (Problem 9.3) we had the reduction potential of oxygen, at pH $= 0$, given as

$$1/2\,O_2 + 4H^+ + 4e^- \rightarrow H_2O \qquad E^{\ominus}\,1.229\,\text{V(pH 0)}$$

The potential 1.229 V is determined with respect to the standard hydrogen electrode, SHE, at pH 0. At another pH the potential of SHE will change according to the formula

$$E(\text{SHE}) = 0.0[\text{V}] - 5.92 \times 10^{-2} \times \text{pH}$$

So at pH 5, which is reported to be optimal for the PSII [5], the potential of SHE – and every other electrode – will have to be lowered by $5 \times 5.92 \times 10^{-2}$ V, and the reduction potential for O_2 will be given as

$$E^{\ominus}(O_2) = 1.229[\text{V}] - 5.92 \times 10^{-2}[\text{V}] \times 5 = 0.933\,\text{V}$$

In the photosynthetic center II exactly the opposite process happens: water gets reduced to give dioxygen. So the potential expected for reaction, at pH $= 5$, will be $E^{\ominus}(H_2O/O_2) = -0.933$ V. The full reaction for each photosynthetic cycle involves two water molecules, so we write

$$2\,H_2O + 4\,e^- \rightarrow O_2 + 4\,H^+$$

and the changes in the standard Gibbs energy will be calculated according to the relation

$$\Delta G^{\ominus} = -zE^{\ominus}F = -4 \times (-0.933[\text{V}]) \times 96485[\text{C mol}^{-1}]$$
$$= +3.601 \times 10^{5}\,\text{J mol}^{-1} = 360.1\,\text{kJ mol}^{-1}$$

Now we need only few key punches to calculate the efficiency:

Efficiency = produced energy/supplied energy

Efficiency = $\Delta G^{\ominus}/E$(light,mole) = 3.601 kJ mol^{-1}/703.7 kJ mol^{-1} = 0.4928
            = 49.28%

The apparent efficiency calculated using this, very simple, scheme is about one-half. This is about right, given that the total efficiency of the PSII and PSI – the photosynthetic system 1, involved in the conversion of H_2O and CO_2 into glucose – is about 25% [3–6]. This may not strike you as nature's greatest design but keep in mind that electrolytic splitting of water is a difficult reaction to carry out in lab, requiring a lot more energy than a few photons harvested by the PSII. We still have few more things to learn from mother nature.

(B) The PSII, like our cars, is a four-stroke (or nearly so) engine: the stage S_0 (which absorbs one photon of light) takes 30 μs to turn into S_1; another photon absorbed and 70 μs later it is S_2; another photon and another 190 μs and it is the stage S_3, and yet another photon and 200 μs gets it to S_4. After the S_4 stage some things happen in the dark (like another stage, S_4) and they take the whole 1.1 ms to reach the beginning, S_0 [5]. What would you say is the rate constant for this engine?

I would solve part (B) by first making this assumption.

» Assumption

• We assume this process occurs according to (pseudo)first-order kinetics.

Pseudo-first order, let us remind ourselves, is when you do not really know the order of the reaction but use so much of the reagents (all but one) that their concentrations are almost constant. Make a note: I said *almost* constant, where almost means there is a change in concentration and there is perhaps an error we are making by assuming the first-order kinetics but the error is small enough that we can ignore it. In practical terms this means I have switched one variable – the true order of the reaction, which I have no way of knowing – to another variable, the concentration of the reagent(s), which I can control. As I said before: a good physical chemist is an artist of approximations and assumptions.

The mathematical form of the first-order kinetics, expressed using the concentration of the product at time t, $P(t)$, and initial product and reactant, $P(0)$ and $R(0)$, is given as

$$P(t) = P_0 + R_0[1 - \exp(-k \times t)] \tag{A}$$

And the reaction rate constant is given as

$$k_1 = -t^{-1} \times \ln\{1 + (P_0/R_0)[1 - P(t)/P_0]\} \tag{B}$$

Now what? We are not given either P_0 or P; however, we do know how much time it takes to split water into dioxygen – it is the time it takes the PSII to make one full cycle, starting from S_0; we will label this time by τ. We calculate τ by adding the times needed to complete one full reaction cycle:

$$\tau = 30\,\mu\,s + 70\,\mu\,s + 190\,\mu\,s + 200\,\mu\,s + 1.1\,ms$$

$$\tau = 30 \times 10^{-6}[s] + 70 \times 10^{-6}[s] + 190 \times 10^{-6}[s] + 200 \times 10^{-6}[s] + 1.1 \times 10^{-3}[s] = 1.59 \times 10^{-3}[s]$$

It takes almost 1.6 ms for PSII complex to form one molecule of dioxygen from water and light. This is the time, let us remind ourselves, needed for the photo-electrochemistry only; the absorption of four photons of light takes "no time." If $\tau = 1.6$ ms we should ask how many dioxygen molecules does the PSII complex churn out in one second; let us call this number n:

$$n = 1[s^{-1}]/1.6 \times 10^{-3}[s] = 629 \approx 630[s^{-1}]$$

Next, we wil use (B) to make an estimate of the rate constant. As we do not have experimental data for R_0 and P_0, we will combine the known information with assumptions. We know the reaction time, $t = 1.59 \times 10^{-3}$s, and we know that it takes two H_2O to make one O_2. Next we will set the initial product concentration to one, $P_0 = 1$, and assume that the reaction course is approximately linear during 10 time periods. We write for (B):

$$k_1 = -(10t)^{-1} \times \ln\{1 + (1/(2 \times 10))[1 - (10/1)]\} = 37.6[s^{-1}]$$

The factor 2 in the denominator accounts for the two water molecules needed for each O_2. Though a crude approximation, k_1 is in the ballpark of the experimental rate constants for PS II [7].

Problem 12.10	A little bit to you and a little bit to me.

The absorption peak of UV light for O_2 molecule is at 145 nm – a deep and harmful UV. As a quantum of this light gets absorbed by O_2 molecule, 15% of the photon energy gets converted to the O–O vibration, $\sigma = 1556.2$ cm^{-1}, and the rest of it is emitted back as light. (A) What is v, the vibrational quantum number of the vibrationally excited O_2 molecule? (B) What is the wavelength of the back-emitted light?

» A and B – Strategy

This problem is about how to divide extra energy in a molecule. It may help us if we break the problem into smaller parts:

1. The O_2 molecule is hit by an energy packet, let us label it as E_1. First, you are going to calculate E_1, the impact energy, using the Planck–de Broglie's formula for energy of electromagnetic radiation and inserting the given value for wavelength; let us label it by λ_1.

2. A smaller part, 15%, of E_1 is immediately converted to the vibrational energy of O_2; we write this as $E_{vib} = 15\%(E_1)$. From the formula for energy of a quantum mechanical oscillator you will calculate v, the quantum number of this vibration.
3. The rest of the energy packet, 85% of it, is bounced back as light, let us label the energy of the returned light as E_2; we write this as $E_2 = 85\%(E_1)$. You are going to calculate how much is $0.85 \times E_1$ and then use again the Planck–de Broglie's formula for the energy of UV light to find out the new wavelength, let us call it λ_2. You have done both types of calculations so – let us see some numbers here, fast.

» **Solution A and B – Calculation**

First, find the energy of the initial (the exciting) light E_1 by using the Planck–de Broglie's formula and inserting the given value for the wavelength, $\lambda_1 = 154$ nm:

$$E_1 = hc/\lambda_1$$

$$E_1 = 6.626 \times 10^{-34}[\text{J s}] \times 2.998 \times 10^8[\text{m s}^{-1}]/145 \times 10^{-9}[\text{m}]$$

$$E_1 = 1.370 \times 10^{-18}\,\text{J}$$

(A) 15% of E_1 is converted (or "lost") to the vibrations of the oxygen atoms in the O_2 molecule. Calculate how much energy is that and equate it with the vibrational energy, E_{vib}:

$$E_{vib} = 0.15 \times E_1$$

$$E_{vib} = 0.15 \times 1.370 \times 10^{-18}\,\text{J} = 2.055 \times 10^{-19}\,\text{J}$$

The expression for the energy of a quantum oscillator is given as

$$E_{vib} = (v + 1/2)hc\sigma$$

Reshuffle the formula to solve it for v, the vibrational quantum number, then insert the value for E_{vib} and calculate v:

$$v = (E_{vib} - 1/2 \times hc\sigma)hc\sigma$$

Evaluating the $hc\sigma$ product, which appears in both the numerator and the denominator, will simplify the final calculation:

$$hc\sigma = 6.626 \times 10^{-34}[\text{J s}] \times 2.998 \times 10^8[\text{m s}^{-1}] \times 1556.2[\text{cm}^{-1}]$$

$$\times 100[\text{m}^{-1}/\text{cm}^{-1}] = 3.091 \times 10^{-20}\,\text{J}$$

Note the cm^{-1} to m^{-1} conversion factor: $100\ [\text{m}^{-1}/\text{cm}^{-1}]$. So, finally, you get for the vibrational quantum number, v

$$v = (2.055 \times 10^{-19}[\text{J}] - 1/2 \times 3.091 \times 10^{-20}[\text{J}])/3.091 \times 10^{-20}[\text{J}] = 6.145 \approx 6$$

After receiving 15% of the initial photon energy the O_2 molecule is vibrating in the 6th vibrational level; we say the molecule is in a vibrationally excited state.

(B) 85% of the initial energy packet is emitted back as light. After this happens the O_2 molecule is not anymore in an electronically excited state; it is, as we just calculated, in a vibrationally excited state. You will first calculate how much energy is emitted back, E_2, and then use the Planck–de Broglie's formula to calculate the wavelength of the light that corresponds to this energy:

$$E_2 = 0.85 \times E_1 = 0.85 \times 1.370 \times 10^{-18}\,\text{J} = 1.164 \times 10^{-18}\,\text{J}$$

$$E_2 = hc/\lambda_2$$

And now solve the latter equation for λ_2:

$$\lambda_2 = hc/E_2 = 6.626 \times 10^{-34}\,[\text{J s}] \times 2.998 \times 10^8\,[\text{m s}^{-1}]/1.164 \times 10^{-18}\,\text{J}$$

$$\lambda_2 = 1.706 \times 10^{-7}\,[\text{m}] = 170.6\,\text{nm}$$

This is the wavelength of the emitted light.

A comment on light and heat: There is a lesson – actually, two lessons – to be learned from these numbers. When the initial light, $\lambda_1 = 145$ nm, impacts the molecule, it causes an electron to jump to a higher level and the molecule is now electronically excited. And what comes after excitation? Relaxation – the molecule relaxes by emitting the light back; light emission is therefore known as *radiative relaxation*. Note the wavelength of the emitted light, $\lambda_2 = 170.6$ nm; this is still a hard UV light but the wavelength is longer. If this process took place in the visible light range and the exciting light was orange the returned light would be red; it would be *red-shifted*. This is a general rule – the emitted light is of lower energy. Since red light is lowest in energy of all visible light we say that the light emitted back by an electronically excited molecule (or atom) is red-shifted compared to the initial light, the light that got it excited. We use this term for very hard cosmic radiation as well as very soft radiowaves.

Not every photon of light, absorbed by a molecule or atom, is emitted back. Sometimes all of the absorbed light is down-converted to smaller energy packets, like atomic vibrations or even smaller, like molecular rotations. We have a generic name for these processes: we call them *non-radiative relaxation*. In other cases only a part of the light energy is converted to vibrations – like in the problem above. The electronically excited O_2 molecule emits back 85% of the initial photon energy and the rest of the energy is used to excite vibrating atoms into 6th level. This is a highly vibrationally excited molecule and, as molecular vibrations are related to heat, we say this molecule is very hot, or it has a *high vibrational temperature*. True, heat and temperature are properties that apply to a large number of molecules, i.e., an Avogadro number, but physical chemists sometimes talk (not quite correctly) about a temperature of a single molecule. After all, it is the energy exchange we have in mind, in either form or under whatever name. To sum it up, this problem is about a rather common process, taking place all the time in the world around us: when a light gets absorbed by a material the emitted light is of smaller energy (red-shifted) and the material heats up through this energy exchange. We can think of this as yet another example of the first law of thermodynamics: no energy is created or destroyed. In case of the interaction of light with matter it just "crumbles" down into smaller energy packets.

Problem 12.11 | One to go.

Your cell phone operates at 860 MHz frequency. Compare a photon of this energy with the following: (A) the average OH bond energy at 336 kJ mol^{-1} and (B) the weakest hydrogen bonding, HB, interaction, with the energy around 8 kJ mol^{-1}. Which of these energies is higher — the cell phone radiofrequency or the covalent and hydrogen bonds? Do you think, on basis of these numbers, that talking on the cell phone could interfere with the chemistry of your brain (assuming no fast driving and talking on the phone is involved)? Support your answer with a number.

» Solution – This one is on you

References

Units and Constants
1. Woan, G (2003) The Cambridge handbook of physics formulas. Cambridge University Press, Cambridge
2. National Institute of Standards URL: http://physics.nist.gov/constants, Accessed July 31, 2009

Photosynthetic System II
3. Ferreira KN, Iverson TM, Maghlaoul K, Barber J, Iwata, S (2004) Architecture of the photosynthetic oxygen-evolving center. Science, 303:1831–1838
4. Kok B, Forbush B, McGloin, M (1970) Cooperation of charges in photosynthetic O_2 evolution – I. A linear four-step mechanism. Photochem Photobiol, 11:457–476
5. Haumann M, Liebisch P, Müller C, Barra M, Grabolle M, Dau H (2005) Photosynthetic O_2 formation tracked by time-resolved x-ray experiments. Science, 310:1019–1021
6. Brudvig GW (2008) Water oxidation chemistry of photosystem II. Phil Trans Roy Soc, B 363:1211–1219
7. Hiller W, Wydrzynski T (2000) The affinities for the two substrate water binding sites in the O_2 evolving complex of photosystem II vary independently during S-state turnover. Biochemistry 39:4399–4405.

Table VI

Summary of molecular spectroscopies – what have we learned in this section?

There are a number of new words and concepts in this section. Find the most important ones and write them down. Write your explanation by each of them. Then make a table of most important formulas

Words and Phrases	Symbols, Formulas and Numbers
Particles and waves	c (speed) $= \nu$ (frequency) $\times \lambda$ (wavelength) c(light) $= 2.99792458 \times 10^{+8}$ m s^{-1}
Planck–de Broglie equation	$E = h\nu = hc/\lambda$ h (Planck constant) $= 6.62606896 \times 10^{-34}$ J s
Pauli exclusion principle Heisenberg Uncertainty Principle	$dx\, dE > h/2\pi$
Infinite potential well – square	$E_n = n^2 \times h^2/8m_e L^2$, $n = 1, 2, 3, \ldots$
Infinite potential well – circular	$E_m = m^2 \times h^2/8\,m_e\,\pi^2\,r^2$, $m = 0, \pm 1, \pm 2, \ldots$
HOMO–LUMO gap in a square well	$\Delta E = [(n+1)^2 - n^2] \times h^2/8m_e L^2$
UV/Vis/NIR spectrum	200 nm $< \lambda <$ 1000 nm, nm $= 10^{-9}$ m
Bouguer–Lambert–Beer law	$A = \varepsilon$(coeff.) $\times l$ (path) $\times c$(mol L^{-1}), $T = 10^{-A}$
Atomic vibrations, energy and units; vibrational levels	$E_v = (1/2 + v)\, h\, c\, \sigma$, σ (wavenumber) [cm^{-1}] v(vibrational quantum number) $= 0, 1, 2, 3, \ldots$
Force constant, k; reduced mass, μ	$\sigma = (2\pi c)^{-1} \sqrt{(k/\mu)}$, $\mu = m_1 m_2/(m_1 + m_2)$
Nuclear magnetization	$\nu_0 = \gamma \times B/2\pi$ [s^{-1}], B(magnetization) [T] γ(atom) $=$ magnetogyric ratio [MHz T^{-1}]
Chemical shift	$\delta \times 10^{-6} = (\nu - \nu_0)/\nu_0$ [ppm]
Boltzmann distribution of energy levels; $\beta > \alpha$	$N_\beta/N_\alpha = \exp[-E/k_B\,T]$ k_B (Boltzmann const.) $= R$(gas const.)/N_A

Index

Note: The locators in boldface refer to definitions cited in the text.

A

absorbance (and transmittance; light, spectrophotometry), 84, 173–**174**
Absorption (extinction) coefficient, **173**–174
Absorption (maximum), 174
Acids (and bases, Brønsted), **87**–88, 100, 105, 117, 144
Activation energy (Arrhenius), 145–149
Activity (and activity coefficient), **61**–62, 93, 103–105, 107–108, 116, 121, 127, 139
Activity coefficient (electrolytes), 63, **103**–105, 107–108, 127
Air, 16–18
Amino acids, 91, 182
Anharmonic, 180
Anion, 90, **94**, 99, 103, 114
Atmosphere (Earth), 39, 194
Atmosphere (pressure unit), **16**, 67–68

B

Balloon (hot-air), 9, 13, **14**–16
Beer (Bouguer-Lambert), **173**–175, 203
Bioelectrochemistry, **117**–118, 127
Bioluminescence, **172**
Boltzmann (constant), 37, **190**–191, 203
Brønsted's acidity, **87**
Buoyancy (cold air, hot air), 14–16, 31

C

Cation, 93, **99**, 103, 106, 108
Cell phone (operational frequency; energy), 117, 155, 166, **201**
Chemical (equilibrium), 11, 66, **68**, 79, 95, 106, 108, 111, 141, 143–144, 177
Chemical potential, **41**, 43, 111, 114
Chemical reaction quotient (Q), 65–67, 75, 95, 120

Chemical

Chemical (shift), **186**–188, 203
Chemiluminescence, 172
Chimpanzee (*Pan troglodytes*, phylogeny), 136–137
Colligative (properties), 100–**102**
Concentration (percent, molal, molar), **53**–54, 56–57, 61–63, 65–67, 73–74, 76–83, 85, 88–89, 93, 95, 103–105, 107, 114–116, 121, 123–126, 132–136, 138, 142–145, 173–176, 190, 197–198
Cooperativity (positive), 76–**77**
Cortisol (and glucocorticoid receptor), **71**, 73, 75, 92, 94
Coulomb (coulombic interaction), 156, **193**
Creatine kinase, **141**, 143

D

Dalton (Law), 17
Daniell (electrochemical cell), **114**, 117–118, 120, 122, 127
De Broglie (Planck – de Broglie), **156**–157, 165, 190, 193, 198–199, 203
Debye (Debye-Hückel Limiting Law, DHLL), **103**–104, 127
Dialysis, **78**, 118, 123
DNA (single & double stranded), **83**–87, 135, 137–138
Donnan (potential), **118**, 123, 125–126

E

Electrochemistry, 111–126
Electrode (potential, standard hydrogen), **117**–118, 196
Electrolytes, 102–103, 127
Electrolytic (cell, reaction), 117, 197
Electronvolt, 193

P.-P. Ilich, *Selected Problems in Physical Chemistry*,
DOI 10.1007/978-3-642-04327-7, © Springer-Verlag Berlin Heidelberg 2010

Endergonic (and exergonic), **64**, 122
Enthalpy (bond), 192, 194
Enthalpy (point function), **33**–34, 44–47, 49, 66, 70–72, 82, 84, 86, 193
Entropy (concept of, change), 35, 38, 40, 44–47, 70, 86
Enzymes (properties, classification, kinetics), 121, **140**–141, 143, 147, 149
Equilibrium (chemical), 71, 79, 95, 143
Excited state (electronic), **164**, 166
Excited state (nuclear magnetic), 184–**185**, 187–191
Excited state (vibrational), **200**
Exergonic (and endergonic), 64
Exhale (expire; inhale), 9, 16–18

F
Faraday (constant), **115**
Fluorescence (phosphorescence), 172
Force constant, 178, **180**, 182
Formal oxidation state, **120**
Frequency (radiowave, infrared, UV/Vis light; and energy), **157**, 179, 185–190, 193, 201, 203

G
Gas constant, 11–12, 19, 61, 149, 190
Gas equation, law, 11, **16**, 18
Gibbs (energy, free energy, standard free energy), 34, 40–45, 47, 49, 61–94, 102–103, 115–116, 119–122, 124, 145–147, 184, 196
Gibbs (Josiah Willard), **40**, 44–45, 47, 49
Glucocorticoid receptor (and cortisol ligand), **71**, 75
Ground electronic state, **163**, 172
Ground nuclear magnetic state, 186
Ground vibrational state, *see* ZPE, zero potential energy

H
Half-cell, 111, **114**–118, 120, 122, 127
Harmonic (motion, pendulum, vibration), **177**–178, 180
Heat (capacity), **23**, 25–26, 28–29, 30, 44, 47, 49
Heat (content, transfer, exchange, flow, gain, loss, absorbed, released), 23–34, 38, 40, 49, 66, 200–201
Heat of vaporization, **28**
Heat (path function), 33
Heisenberg (uncertainty principle), **177**, 203
Hemoglobin (ligand:receptor interaction, equilibrium), 18, 76–**77**, 166
Henderson-Hasselbalch (acid-base equilibrium), **88**, 95

HOMO-LUMO, 161, **163**–168, 171, 203
Homo sapiens (phylogeny), 135
Hückel (Debye-Hückel Limiting Law, DHLL), **103**–104, 127

I
Inhale (inspire; exhale), 16–18, 122
Internal energy, 33–34, 49
Ionic strength, 104–107, 127
Ions (and ionic properties), 34, 93, 99–108, 111–116, 118, 124–127, 133
Isotopic effect (atomic vibrations), 180–183

K
Kinesin, 4–5
Kinetics (except enzyme kinetics), 131–149, 195, 197

L
Ligand (and receptor), *see* Receptor (and ligand)
Light, light & matter interaction of, 47, 64, 84, 145, 155, 157, 160–161, 163–166, 168, 171–201
LITTLE BIG TRICK # 1, 12, 183
LITTLE BIG TRICK # 2, 26, 81
LITTLE BIG TRICK # 3, 67, 84, 89, 198
LITTLE BIG TRICK # 4, 39, 67, 74
LITTLE BIG TRICK # 5, 69, 87
Liver alcohol dehydrogenase, LADH, 119
Luminescence (fluorescence, phosphorescence), 171
LUMO (HOMO), 161, **163**–166, 168, 171, 192, 203

M
Macroscopic (microscopic, mesoscopic), 155–156, 158
Magnet (nuclear), *see* NMR (nuclear magnetic spectroscopy)
Membrane (semi-permeable), 77–78, 118, 123, 127
Michaelis (Menten; enzyme kinetics), 141–142, **144**–145, 151
Mitochondrial clock, 135, 137–138, 186
Mixing (entropy of), **39**–40
Molar volume (partial), **55**, 57–**58**, 95

N
NADP/NADPH (nicotineamide dinucleotide phosphate), **119**–123, 127
Neanderthal (*Homo neanderthalensis*, phylogeny), **135**, 137–138
Nernst's equation, **116**–117, 122, 127
Newton, 3–4, 19

Newtonian (mechanics), 156
NMR (nuclear magnetic spectroscopy),
    184–191

O
Oxidation, 66, 111, **118**–121, 147
Oxygen, dioxygen (emergence; production),
    18, 34, 39, 53, 76, 118, 122, 194
Oxygen, dioxygen (reduction potential), 118,
    122–123, 160, 195–196

P
Pascal (air pressure), **19**, 42, 68
Pathlength (optical cell), 174–176
Pendulum (harmonic & anharmonic),
    177–178, 180
pH (and standard hydrogen electrode
    potential), **117**–118, 196
pH (concentration, activity of H^+), **93**, 95,
    104–105, 107, 121, 123
Phosphorescence (luminescence), 172
Photochemistry, 192
Photoionization, 192
Photon (energy), 155, 161, 163–166, 168, 171,
    188–189, 192–198, 200–201
Photosynthesis (Photosynthetic system II & I),
    194–198
Planck – de B roglie, **156**–157, 165, 190, 193,
    198–200
Planck, Max (constant), **156**–157, 163, 165,
    177, 190
Plutonium (decay: kinetics, energy,
    pacemaker), **138**
Population (energy levels), 135, 189, 191–192
Porphin, porphyrin (conjugation, light
    absorption), 166–168, 174
Potential (energy), 156–157, 177–178
Potential (well, infinite; linear, circular),
    157–162, **167**, 203
Pressure (gas, ice-skate), 9, **11**, 15–19, 49, 61,
    66–68, 76
Pressure (skate-blade), 40–42
Probability (& entropy), 37–38, 49
Proline (atomic vibrations, isotopic
    substitution), 181–183
Pseudo (first order kinetics), 3, 144, 148, **197**
Psi (pound per square inch – pressure unit),
    9–11, 13

Q
Q (reaction quotient), **65**–66, 95, 115–116,
    120–121
Quantum (quantum mechanics), 145, 156–**157**

R
Radioactivity (Pu-238), 138–140
Receptor (and ligand), 62, **68**–83, 94, 97, 138,
    141, 199, 201
Red-shift, 200–201
Reduced mass, **178**, 181–183
Reduction (reductive potential), 115, 117–120,
    122, 126–**127**, 195–196
Relaxation (non-radiative, radiative), 172, 185,
    200
Retinal (conjugation, vision, HOMO-LUMO),
    161, 163–166, 168

S
Skating (ice, water, Gibbs free energy), **40**, 41,
    43, 64, 100
Solute (solvent interaction), **53**–56, 61–62,
    95, 100
Solution(s), 53, 99–108, 111–117, 119, 122,
    125–127, 144, 174–176, 182
Solvent (volume contraction), 53–58, 67, 95,
    100, 143
Spectrophotometry, 84, 172, 174–176
Spectroscopy, 171, 176, 181, 184, 186–188,
    190–191

T
Temperature (body), 16, 18, 23, 25–29, 32, 35,
    37–38, 40, 44–47, 71, 77, 86, 147–148
Temperature (vibrational), 178–179, **200**
Thermochemistry, **34**
Thermodynamic, First Law, 35, 40, 44, 201
Thermodynamics, 71, 95, 145, 164–165, 195
Thermodynamic, Second Law, 35, 40
Torque, 6–**7**, 19
Tryptophan (acid-base equilibria), **91**, 93, 95,
    106, 108
Tyrosine (nitration), 186–188

V
Vaporization (heat of), **28**
Vibrational spectroscopy, 176, 181, 192
Vibration (oscillation, atomic), **177**–179, 199,
    200
Vision (retinal), **161**
Volume (molar, partial), **43**, 57–**58**, 95

W
Wavelength (light), 157, 160–161, 163,
    165–166, **168**, 171, 173–174, 189, 192–193,
    198–200, 203

Wavelength (particle), 156
Wave mechanics, 156–157
Wavenumber (vibrational spectroscopy),
177–182, 203
Work (electrical), 118
Work (mechanical), 3–4, 13, 15, 23, 31, 33–34,
44, 49, 61, 184

X
Xanthine (oxidase: hypoxantine, uric acid),
147–148

Z
ZPE, zero potential energy, **177**
Zwitterion (tryptophan), **94**